Solar Flare Survival

Marc Remillard

Printed in the United States of America

Cover Image Courtesy of NASA

Solar Flare Survival

Marc Remillard

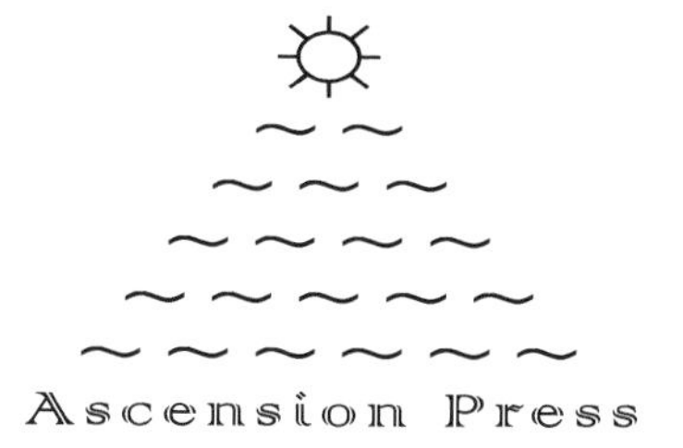

Table of Contents

Watching The Sun

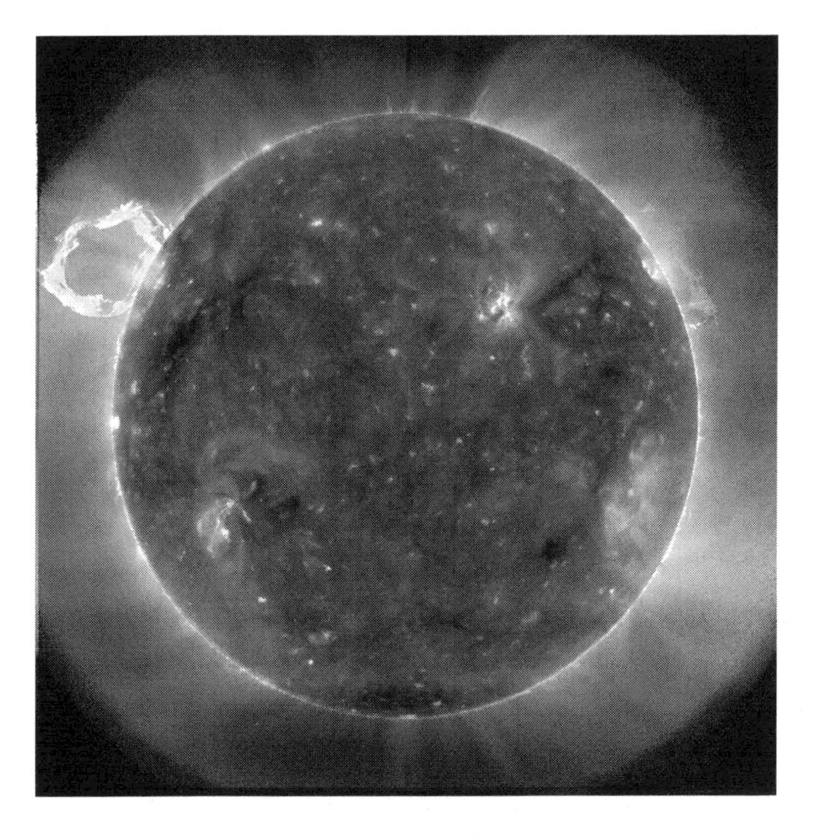

Large Coronal Loop
Image Courtesy of NASA

Definition of Terms

Sunspots

Sunspots are dark areas that form and disappear on the surface of the sun over periods of days or weeks. Sunspots are caused by concentrated magnetic fields that reduce the amount of energy flow to the surface of the sun from its interior. The reduced energy flow causes the area to cool from about 5,800° C to 4,300° C. Because sunspots are cooler than the rest of the Sun, they appear dark on the Sun's surface. Sunspots are so big that all of planet Earth would easily fit into a sunspot. They start out as magnetic knots in the solar convection zone, which is the outermost layer of the Sun's body and also the region likeliest to be disturbed by external electromagnetic or gravitational influences. Currents of plasma, or highly electrified gas, act as conveyor belts and ferry these knots from the poles to the equator, where they rise up to the surface and explode as magnetic storms.

Solar Cycles

About every 11 years the sun flips with respect to its magnetic field. Solar cycles as short as 9 years and as long as 14 years have been observed, but the average is actually 10.66 years. During each of these periods, scientists observe an increase in sunspot activity called a *solar maximum,* followed by a period of lesser sunspot activity called a *solar minimum.*

Solar Flares

Solar flares are the release of a single burst of energy from the sun, which may take any of many electromagnetic forms. The forms vary from radio waves through the visible spectrum to gamma rays and x-rays, energetic particles (protons and electrons), and matter so hot that it is in the form of plasma. Solar flares are characterized by their brightness in x-rays. Most flares occur in active regions around sunspots, where intense magnetic fields penetrate the photosphere to link the corona to the solar interior.

Interplanetary Magnetic Field (IMF)
The solar magnetic field carried by the solar wind among the planets of the Solar System.

Sudden Ionospheric Disturbance (SID)
Sudden ionospheric disturbances are characterized by abnormally high ionization/plasma density in the ionosphere caused by a solar flare. When a solar flare occurs on the Sun a blast of intense ultraviolet and x-ray radiation traveling at the speed of light hits the dayside of the Earth--typically in about eight minutes. SID's can cause radio blackouts, and are the most severe in the equatorial regions where the Sun is directly overhead. Scientists monitor SID's and use them as precursors to alert us to much larger (but slower) CME events.

Coronal Mass Ejections (CME)
Coronal mass ejections are the sudden release of large masses of plasma from the super hot corona, which is the atmosphere just above the surface of the sun. CME's expand away from the Sun at speeds as high as 4 million miles per hour. The mass of particles may take from two to five days to arrive into Earth's arena. Coronal mass ejections are more likely to have a significant effect on our activities than solar flares because they carry more material into a larger volume of interplanetary space, increasing the likelihood that they will interact with the Earth. CME's typically drive shock waves that produce energetic particles that can be damaging to both electronic equipment and astronauts that venture outside the protection of the Earth's magnetic field.

Each category for x-ray flares has nine subdivisions ranging from, e.g., C1 to C9, M1 to M9, and X1 to X+.

Most CME's occur during the solar maximum of the 11 year solar cycle. However, powerful CME events have been recorded during solar minimum periods as well.

Solar Energetic Particles (SEP)
Solar energetic particles are high-energy particles coming from the Sun consisting of protons, electrons and heavy ions. They can originate from either solar flare sites or manifest as shock waves associated with

coronal mass ejections (CME's). SEPs are of particular interest because it is these particles that interact with the Earth's magnetic field, causing damage to equipment and threatens human life.

Magnetosphere

A magnetosphere is formed around the Earth when a blast of particles from a solar energetic particle (SEP) event interacts with the Earth's magnetic field.

The magnetosphere extends out to a distance of the order of ten Earth radii, and is distinctly non-spherical. It is in the form of an oval tear-drop shape because of effects of the solar wind.

Electrojets

When solar energetic particles (SEP's) or a coronal mass ejection (CME) form a magnetosphere, energetic particles can flow in through the Earth's polar regions and form electrojets. Most prevalent in northern and southern latitudes, electrojets swirl around the Earth and touch down on the surface in continent-wide footprints. Electrojets are unpredictable and change polarities constantly.

Geomagnetically Induced Currents (GIC)

When electrojets contact the surface of the Earth, they create geomagnetically induced currents (GIC's). These powerful currents are generally on the order of 10's to 100's of amperes, and can flow through power lines, conductive seawater and underneath the surface of the Earth. Many GIC'c flow 400 kilometers beneath Earths surface, but some are as deep as 700 kilometers. GIC's have a documented history of destroying large power transformers used in national power grid systems.

Faraday Cage

An enclosure designed to provide a shield to prevent electromagnetic energy from either penetrating, or escaping.

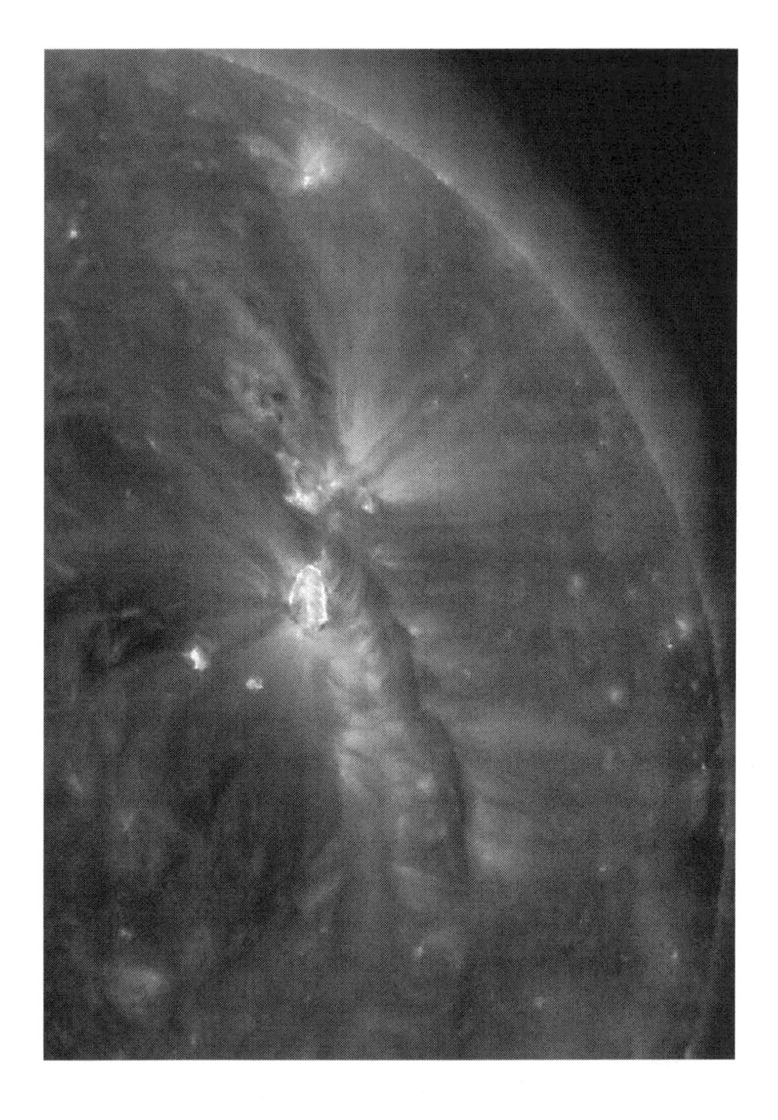

Multiple Solar Flares
Courtesy of NASA

Introduction

As modern technology marches boldly forward, humankind is experiencing freedoms undreamed of a mere hundred years ago. The digital age has literally exploded in the last 20 years, with a quasar effect that will certainly be noted by historians far into the distant future.

With that freedom unfortunately comes vulnerability, the extent of which may be remembered as the Achilles heel of the early 21st century. We are akin to a prize-fighter who has become so filled with himself because of a recent winning streak that he foolishly leaves his guard down when faced with a powerful and relentless foe.

Let us, for one moment address the situation courageously by asking two questions that may be both uncomfortable and revealing:

What if most of the electronic devices on our planet were disabled, as well as the machinery used to make those devices--*and* the tools used to make the machinery?

How long would it take for our society to recover from a crippling blow such as that?

We will not spend time tediously endeavoring to answer those questions, nor will we lapse into harping about the ills of our society.

What we shall first discuss is the basics of what our scientists so far have learned concerning the nature of extreme solar events, and how they effect our world. Then we will examine ways to protect ourselves and our electronic gear from potentially crippling solar weather.

We would like to point out that on every corner of our planet there exist very ingenious people; men and women who live and breath to creatively solve problems. Our history underscores this, and it is perhaps our strongest card in this game of Russian Roulette we are playing with our Sun as we approach the apex of the current solar cycle.

This is not a doom & gloom book. There is already plenty of that flavor of information available for those that are so inclined. In fact, we would be delusional to assume that we could even begin to compete in such exotic realms. Nor will we endeavor to capitalize on the disaster hysteria induced by the potential of cataclysmic natural events. We must however, be realistic and examine the facts with an open mind.

> The most powerful solar flare in years last week served as a reminder of the dangers of solar storms to a society increasingly dependent on electronics and communications. (credit: NASA)

The simple statement from NASA sums it up. They were referring to the powerful solar event that occurred on February 14th, 2011, when an class X solar flare erupted. The solar flare was followed by a series of coronal mass ejections (CMEs), which blasted charged particles towards the Earth.

Fortunately, the direst predictions failed to come to pass, as the solar flare caused only minor interference with communications. However, the bottom line is; most of the magnetic storm barely missed us, and dissipated off into space.

Meanwhile, the news media "talking heads" treated the potentially dangerous situation as if it were just another sporting event. Their smug, cavalier attitude was both disturbing and adolescent.

The solar eruption *almost* made headline news.

We were also amused by the pathetic attempts to placate and reassure the populace.

From ABC World News:

> "Do not be afraid. NOAA's Space Weather Prediction Center in Boulder, Colorado, says the Earth is well-protected by its atmosphere and magnetic field."

OK. That's like saying, "Don't worry--*that* Grizzly bear may not be hungry right now!"

In that intellectual climate the idea for this book was born. It sprung from the need to protect ourselves and our electrical equipment from damage or destruction by powerful solar activity. We have no intention to be victimized by a natural occurrence. The idea to write this book came from a desire to protect my own family from the ramifications of extreme space weather.

After performing an internet search, I found the available information on Faraday cage design to be repetitive and contradictory. I was left with some elementary, but important questions. It was actually those questions that gave me the idea to pull together all of the information, examine it clearly, and put together this work.

This book is written for everyone, irregardless of belief systems or scientific leanings. I made a conscious choice to use conservative, well-accepted science as a basis for the ideas expressed here. The discussions on astrophysics, heliophysics, and seismology are boiled down to roughly the easy-to-read 8th grade level. The goal is to create protective systems for ourselves and our equipment, but first we need to understand why it is so important to do so.

However, as I was doing the research for this book, I encountered many remarkable new developments, research, and ideas fermenting in the scientific community. New research papers are published at such rapidity that any rendering such as this book makes it inevitably destined for obsolescence. It would take a large, well managed reporting team to keep up with the research being done by universities, companies, organizations, and individuals world-wide.

Personally, I lean strongly towards cutting-edge science, some of which may (when first proposed) be deemed "pseudo science".

Historically, harsh judgements that are directed towards any new developments tend to be subjective in nature, and driven by the less--than-noble human issues we are all too familiar with. Correct or not, somehow the visionary that steps into the sun is inevitably attacked by anti-bodies. The tendency to resist new ideas was described so clearly by a blogger that we feel compelled to quote him.

From Discovery.com's comment forum:

"One of the biggest assumptions of science is that conditions are always the same, that because things occur right now at a certain rate, and have never been reported to occur at any different rate, that this rate must be fixed and extend to it's logical conclusion in the distant past and/or future. They have applied this to the movement of continents, to the evolution of species, to the development of stars, planets, and ecosystems, and it is nothing but a big fat ugly assumption, that the majority of people simply believe without question.

Use your brain. Recognize that conditions change, or at least that conditions COULD change in such as a way as to create a much different picture of this universe than that which is assumed by the majority of scientists. In an endeavor that is meant to discover reality by empirical investigation, the extension of ideas gleaned from relatively limited data into ever-present truths is of the greatest danger to the genuineness of the enterprise as a whole. At it's very most, all that science has discovered is only that which could be uncovered and observed within the span of a few hundred years, and any assessment of their conditions (and more importantly, resistance to new evidence by the establishment) should keep this limitation at the forefront of one's mind."

Riley Dinsmore

Well said Riley!

Mr. Dinsmore was commenting in a Discovery News blog titled: *Is The Sun Emitting a Mystery Particle* by Ian O'Neill.

The discussion concerned recently observed inconsistencies in radioactive decay rate data.

Researchers at Stanford and Purdue Universities have noticed that the decay rates for radioactive elements are *changing*. We are talking about elements with historically "constant" decay rates -- like carbon-14 and manganese-54. Their values aren't supposed to change, which has been in the past the holy grail of carbon-14 dating for organic material associated with modern archaeology.

They found that the decay rates vary cyclically every 33 days - a period of time that matches the rotational period of the core of the Sun.

Coincidence? Peter Sturrock, Stanford professor emeritus of applied physics had a hunch that solar neutrinos might hold the key to this mystery.

The solar core is the source of solar neutrinos, but neutrinos are not supposed to work like that. It all points to the idea that *something* is coming out of the Sun that may be cyclically altering decay rates.

I find new developments fascinating. Is it not one of the most admirable facets of our human consciousness--to wonder, explore, learn and grow?

We have much to learn, and can only vaguely speculate as to how to protect ourselves from exotic particles such as neutrinos, the elusive graviton, or other entities of physics as yet to be discovered.

That said, I must reiterate that I am using orthodox science as the basis for this text. Occasionally however, I may vacillate by putting a little toe into the fast moving stream.

When we make observations, we should endeavor not to allow them to be tempered by our needs or previous conceptions.

Pure observation may be easier said than done, especially considering the possibility of the controversial *observer effect* entertained by quantum mechanics proponents.

Here our focus is directed towards how to assemble functional, simple Faraday cages systems. This mundane task may not necessarily require cutting edge physics, but we must remember that there remains a good measure of unexplained phenomenon associated with electromagnetic energy.

The morning of August 9th 2011, at 0805 UT, sunspot 1263 produced an X7-class solar flare--so far the most powerful flare of new Solar Cycle 24. The brunt of the explosion was not Earth directed. Even so, a minor solar proton storm enveloped our planet, and the flare also created waves of ionization in Earth's upper atmosphere.

Less than a month later on Sept. 6th 2011, active sunspot 1283 produced two major eruptions including an impulsive X2-class solar flare. The blasts hurled a pair of coronal mass ejections (CMEs) toward Earth. The "X" designation stands for "extreme", while M-class refers to a "moderate" storm (but still powerful).

As we approach both the apexes of the 11 year solar cycle *and* the longer Grand Cycle the frequency of solar storms is increasing. So far in late 2011 our sun has produced five X-class flares, which is unprecedented in our recent terrestrial history.

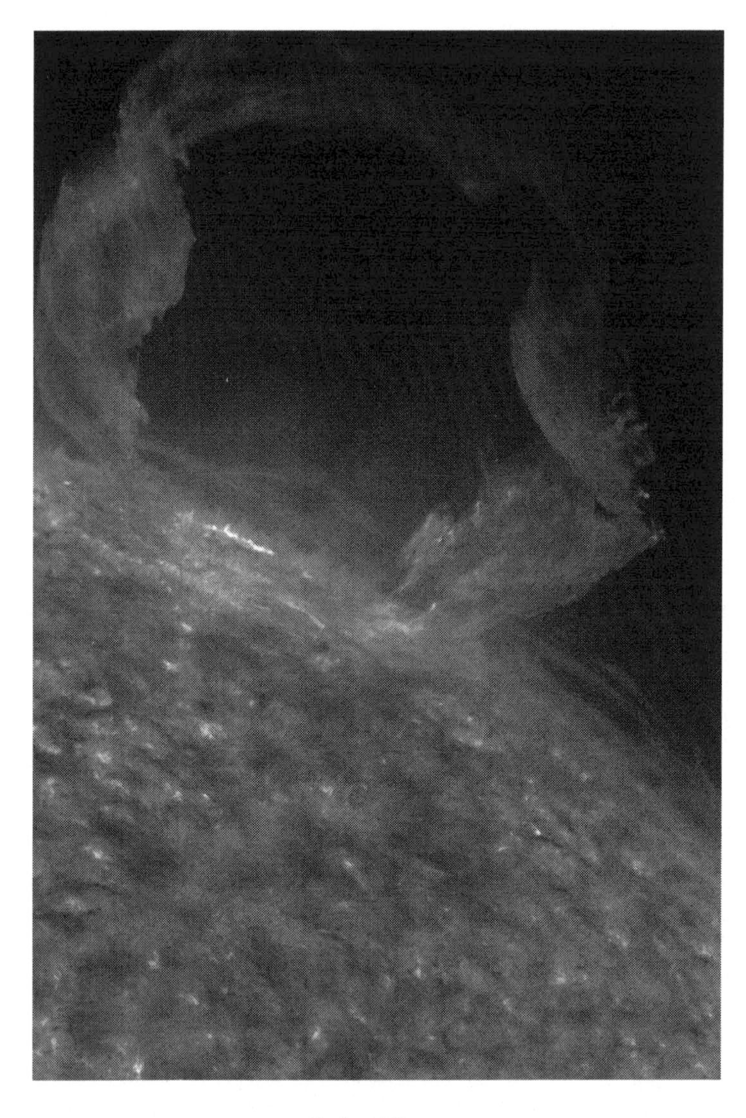

Solar Flare
Courtesy of NASA

Solar Flares

Since the formation of our planet billions of years ago, the Earth has evolved in a highly radioactive environment. The Sun continually fires high-energy particles from its magnetically dominated surface. The flow of particles is known as the *solar wind.* When the Sun is at its most active, (during solar maximum) the Earth may be unlucky enough to be the target of an explosion emanating from the sun with the energy of 100 billion atomic bombs. This is known as a *solar flare.*

Solar flares are classified as A, B, C, M, X and X+ according to their intensity. "A" class flares are the weakest, while "X+" class are the most powerful. Each category has nine subdivisions ranging from C1 to C9, M1 to M9, and X1 to X9, while X+ class are X10 and above.

NOAA has simplified the X-class for their purposes by referring to X class flares from 1 to 20. That is--X1 through X20. NOAA's classification system will most likely become the standard.

M class flares can result in temporary radio blackouts on Earth, but it is the X class events that we are primarily concerned with.

Solar flares affect all layers of the solar atmosphere, heating plasma to tens of millions of kelvins and accelerating electrons, protons, and heavier ions to near the speed of light. Most flares occur in active regions around sunspots, where intense magnetic fields penetrate the photosphere and link the corona to the solar interior.

We are learning more about the nature of our Sun everyday. The current view is that extremes in solar activity are enhanced by the Sun's "natural cycle', a period of approximately 11 years.

During each cycle, the magnetic field lines of the Sun are dragged around the solar body by rotation at the solar equator. The equator is spinning faster than the magnetic poles. Solar plasma drags the magnetic field lines around the Sun, causing a build-up of energy. Eventually, kinks in the magnetic flux form, forcing them to the surface. The kinks are visible and frequent during periods of high solar activity, and are known to us as *coronal loops*. The coronal loops are associated with sunspot activity. This is when a phenomenon called magnetic reconnection occurs.

Reconnection is considered to be the trigger for solar flares large and small. These can be very energetic events, encouraging the acceleration of solar plasma. Solar plasma is superheated particles such as protons, electrons, and light elements such as helium nuclei. In certain conditions, the interaction of solar plasma can cause X-ray emissions, resulting in an X-ray flare.

X-rays travel at almost the speed of light, so Earthlings receive very short warnings when X-ray flares occur. An X-ray from a powerful solar flare can reach Earth in about eight minutes. The resulting disturbances in Earth's ionosphere are known as *Sudden Ionospheric Disturbances* (SID's).

In certain conditions, a *Coronal Mass Ejection* (CME) may occur either at the site of the flare, or independently. CME events travel slower than X-rays, but can reach Earth in as little as thirty minutes, and as much as 4 days. CME's may travel slower, but their effects can be much more dramatic than those from SID X-ray emissions. As we shall see, CME's can cause a considerable amount of bedlam here on Earth.

After a CME occurs on the Sun it pushes against the solar wind, a sphere of charged particles--mostly protons. This causes what is known as a *Solar Energetic Particle* (SEP) event. When the supercharged particles reach Earth, they fuse oxygen and nitrogen atoms to create nitrates, which eventually settle as dust onto the poles.

When the supercharged particles reach Earth, the interplanetary magnetic field (IMF) and the Earths magnetosphere interact. Oftentimes a coronal mass ejection is sluffed off by the magnetosphere, but if the magnetic field line's polarities happen to be in opposite directions reconnection can occur, connecting the Earths magnetic field with the magnetic field of the Sun.

High energy particles then begin flowing into the magnetosphere. Because of the pressure from solar winds, the Sun's magnetic field lines fold around the Earth, and sweep behind the planet. Particles flowing into the "dayside" funnel into the Earths polar regions where they interact with the atmosphere. The visible results of those kind of events are what we know of as the *Aurora.*

High-latitude regions are affected more dramatically when this occurs. A powerful electric current known as an *electrojet* may form, pushing through the atmosphere and bringing with it a magnetic field. Electrojets use our ionosphere as a conduit, flowing to different parts of the globe. When the electromagnetic current contacts the surface of the planet, it can flow through the oceans, and create subterranean flows of energy.

This is NASA's take on it:

> The outer solar atmosphere, the coronal, is structured by strong magnetic fields. Where these fields are closed, often above sunspot groups, the confined solar atmosphere can suddenly and violently release bubbles of gas and magnetic fields called coronal mass ejections. A large CME can contain a billion tons of matter that can be accelerated to several million miles per hour in a spectacular explosion. Solar material streams out through the interplanetary medium, impacting any planet or spacecraft in its path. CMEs are sometimes associated with flares but can occur independently. These are phenomenon that can disrupt power grids and destroy delicate electrical devices.

CME's may travel as fast as four million miles per hour and can carry up to 50 billion tons of plasma.

Solar flares may be observed near the birthplaces of CME's, but solar physicists do not believe that they are what actually cause them. CME plasmas are typically spawned at higher altitudes on the Sun than solar flares. However, the same underlying conditions might be mother to them both.

Recent studies have shown how CME's can actually change directions due to the influence of the Sun's magnetic field.

Jason Byrne and Peter Gallagher of Trinity College in Dublin, Ire-

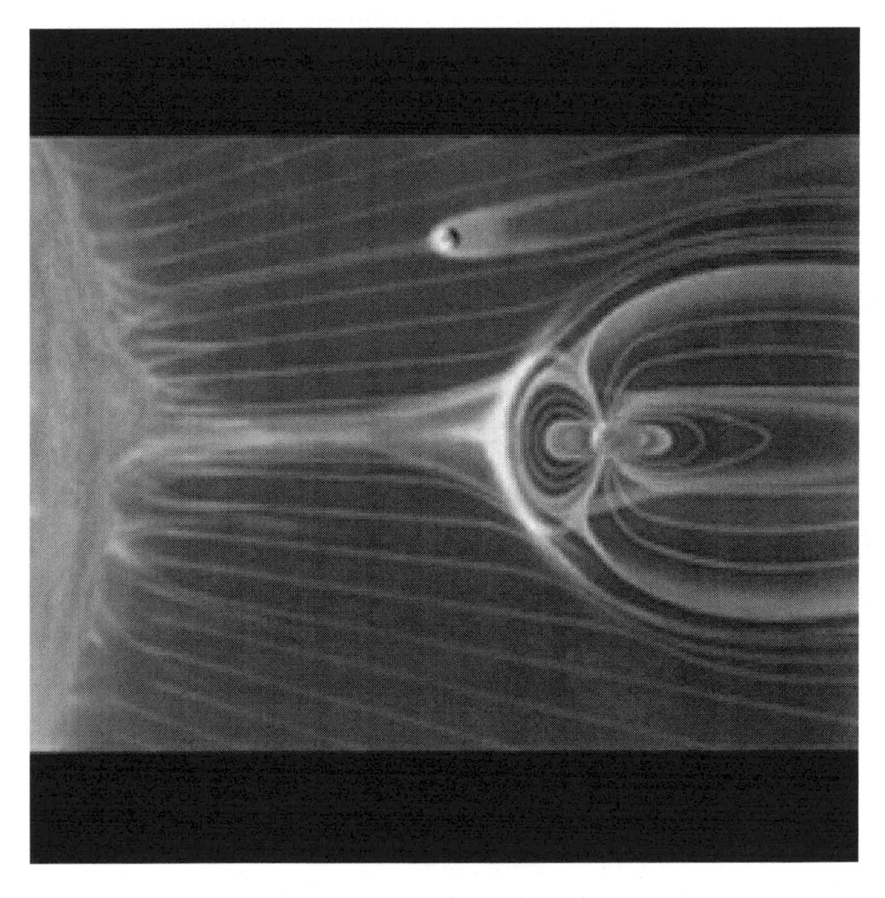

Magnetospheres of Earth and Venus
Artists rendition
Courtesy of NASA

land used an innovative computing technique called *multiscale image processing* to observe coronal mass ejections as they leave the Sun.

Multiscale imaging has been regularly used in the medical industry for diagnostic purposes.

When they applied the multiscale technique to chronograph data from NASA's twin STEREO spacecraft, they were so surprised at the results that they back-tracked and triple-checked their readings. The CME's were not moving in straight lines. Previous studies have suggested the possibility, but this was the first time that they were observed doing so.

The two solar-watching STEREO spacecraft are widely separated and can see CMEs from different points of view. This allowed the team to create fully-stereoscopic models of the storm clouds and track them as they billowed away from the sun.

When a CME ejects out of a northern latitude, the Sun's global magnetic field, which is shaped like a bar magnet, guides the wayward CMEs back toward the sun's equator. This compresses the energy, causing it to accelerate as it blasts out of the Suns immediate magnetic field.

Once a CME is embedded in the solar wind, it can experience significant acceleration. "This is a result of aerodynamic drag," says Byrne. "If the wind is blowing fast enough, it drags the CME along with it—something we actually observed in the STEREO data."

Gallagher's and Byrne's observations are important because they give us more insight into the phenomenal nature of the speed at which CME's travel through the solar system.

Alex Young, STEREO Senior Scientist at the Goddard Space Flight Center commented on the discovery:

> The ability to reconstruct the path of a solar storm through space could be of great benefit to forecasters of space weather at Earth. Knowing when a CME will arrive is crucial for predicting the onset of geomagnetic storms.

Our primary concern is how space weather effects Earth's magnetic field, which has been termed the *geomagnetic field.*

The intricacies of how high energy particles are generated during geomagnetic storms constitute an entire discipline of space science in its own right.

Solar Storm
Courtesy of NASA

For a real-time view of geomagnetic activity over the poles, go online to NOAA/Space Weather Prediction Center > space weather now > auroral map. The auroral maps are centered on the north or south poles, and superimposed over continents in either the northern or southern hemispheres. Check out how vulnerable North America is.

One of the tools used to measure the disturbance of the geomagnetic field is an instrument called a magnetometer. Data is gathered by NOAA from dozens of observatories in one minute intervals. The data is then boiled down for us into a palatable form.

The most popular index used for giving us a quick reading as to the level of geomagnetic intensity reaching the Earth is the *Kp* index devised by Julius Bartels in 1932. Kp lets us monitor how vigorous the Earth's geomagnetic response is to solar activity.

The Kp index is semi-logarithmic, and is from 1 to 10 . A Kp = 9 storm is roughly five times stronger than a Kp = 6 storm. On an ordinary day Earth's magnetic field weighs in from about 1.0 to 3.0 Kp. Kp values between 4.5 and 5.5 are classified as small storms, while storms between 5.5 and 7.5 Kp are considered large. Events eliciting values greater than 7.5 can result in obvious damage, like the 9.3 Kp event that induced the Quebec blackout in 1989.

An even simpler index that is often used to describe the strength of a geomagnetic storm is the "G" scale. How powerful is that geomagnetic event on a scale of one to five?

Scale	Level	Impact	Kp	Frequency
G1	Minor	Radio interference	5	1500 per cycle
G2	Moderate	Power grid disturbances	6	500 per cycle
G3	Strong	Some grid problems	7	100 per cycle
G4	Severe	Transformer trips	8	50 per cycle
G5	Extreme	Blackouts	9	2 per cycle

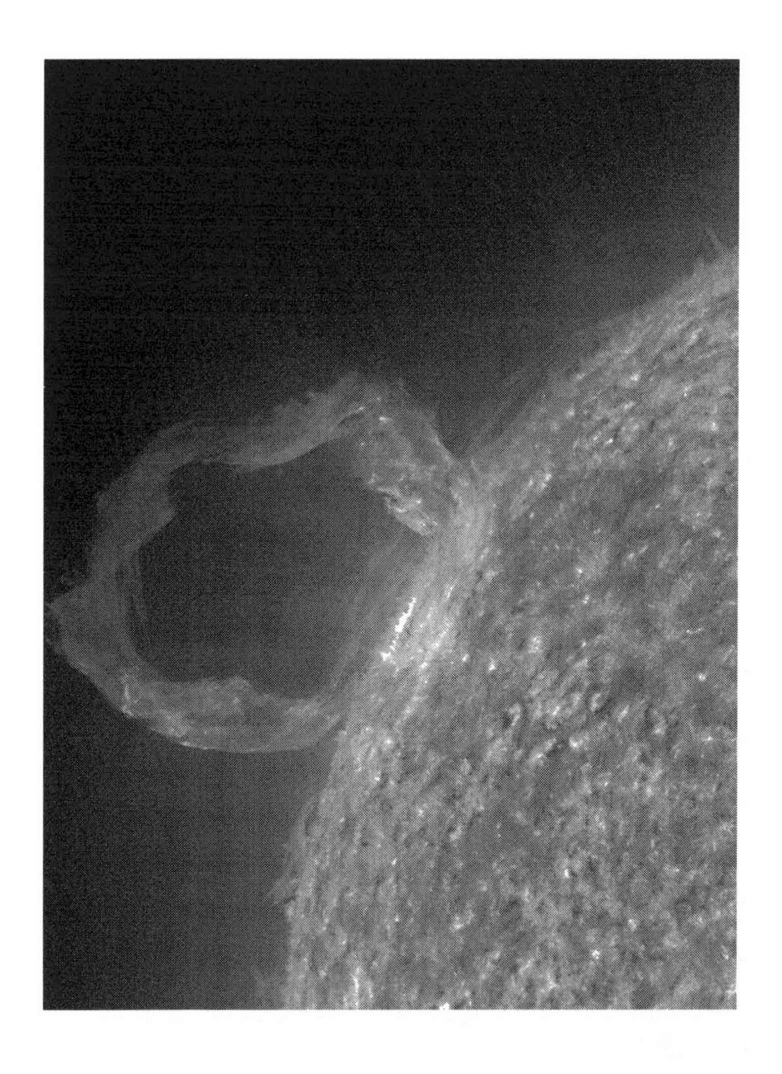

Coronal loop from the SDO spacecraft
Courtesy of NASA

Our Solar Past

Sunspots and solar flares have been observed in the west since the invention of the telescope in 1610. Chinese astronomers have been counting sunspots for over 2000 years. Due to a few early astronomers' fascination with the sun, we have 400 years of accumulated data concerning all facets of solar activity. These records include solar flares, sunspots, long-term solar changes, and reveal the 11 year solar activity cycles.

Our sun also goes through longer cycles, called *grand solar cycles.* According to researchers at Reading University, the current phase began in the 1920's and appears to be peaking now.

Mike Lockwood, professor of space environment physics at Reading comments: "In a grand solar maximum, the peaks of the 11-year sunspot cycle are larger and the average number of solar flares and associated events such as coronal mass ejections are greater."

The research indicates that most radiation strikes the Earth during periods of middling solar activity. We can look forward to an unfortunate combination of solar conditions coming our way in the next few decades as the grand solar cycle begins to wane.

The research is based on evidence from ice core samples and tree rings going back 10,000 years. A research team measured levels of nitrates and cosmogenic isotopes which enter our atmosphere and are deposited in ice and organic material. This gives us a clear picture of the history of SEP (solar energetic particle) events.

One of the longer term cycles is illustrated in the notable period of almost non-existent sunspot activity between 1645 and 1715. This period is known as the "Maunder Minimum," named after the British astronomer Edward W. Maunder (1851–1928), who studied how sunspot latitudes changed with time.

The Maunder Minimum period has also been called the "Little Ice Age" because it was also a prolonged period of low temperatures worldwide. These kinds of events have occurred many times in the geologic past, as evidenced by tree-ring measurements that show slower growth during periods of prolonged cold.

Scientists working with the Soho (Solar Heliospheric Observatory) satellite may have obtained evidence about how the Sun makes the transition from one 11 year cycle to another.

After observing eight years of CMEs - it seems that they are removing the Sun's old magnetic field bit by bit, first from one pole and the equator, and then the other pole.

Nat Gopalswamy of Nasa's Goddard Space Flight Center, author of a report in the Astrophysical Journal states:

> "The Sun is like a snake that sheds its skin. In this case, it's a magnetic skin. The process is long, drawn-out and it's pretty violent. More than a thousand coronal mass ejections, each carrying billions of tons of gas from the polar regions, are needed to clear the old magnetism away. But when it's all over the Sun's magnetic stripes are running in the opposite direction."

Joseph Gurman, Nasa project scientist for Soho comments:

> "This analysis of nearly eight years of CME data is a big step forward in making sense of space weather. "By identifying the solar origin of these events with CMEs of different speeds and appearances, and at different latitudes, it improves our capability to predict space weather that can affect the Earth, at different phases of the solar activity cycle."

The most famous, and certainly one of the most dramatic solar events, occurred on September 1st, 1859, and is called the "Carrington Event". The Carrington Event was the most powerful solar storm in recorded history.

British amateur astronomers Richard Carrington and Richard Hodgson independently made the first observations of a solar flare.

For five days beginning on August 28, 1859, numerous sunspots and solar flares were observed on the sun. Shortly before noon on September 1, the 33 year old English astronomer Richard Carrington observed the largest flare, which caused a huge coronal mass ejection (CME) to travel directly toward Earth. The CME reached Earth in only 18 hours, which is remarkable because such a journey normally takes three to five days. Apparently it moved at that speed because an earlier, smaller CME had created a magnetic field that cleared a pathway for it.

The resulting phenomenon was dramatic.

The new telegraph systems all over Europe and North America failed, and in some cases even shocked telegraph operators. Telegraph pylons threw sparks while telegraph paper spontaneously caught fire. Some telegraph systems acquired phantom power and continued to send and receive messages despite having been disconnected from their power supplies!

There are many interesting, anecdotal stories that were well publicized at the time. One account described the auroral light being so bright over the Rocky Mountains that the glow awoke gold miners, who began preparing breakfast because they thought that it was daybreak.

On September 3rd, 1859, the *Baltimore American and Commercial Advertiser* reported:

> Those who happened to be out late on Thursday night had an opportunity of witnessing another magnificent display of the auroral lights. The phenomena was very similar to the display on Sunday night, though at times the light was, if possible, more brilliant, and the prismatic hues more varied and gorgeous. The light appeared to cover the whole firmament, apparently like a luminous cloud, through which the stars of the larger magnitude indistinctly shone. The light was greater than that of the moon at its full, but had an indescribable softness and delicacy that seemed to envelop everything upon which it rested. Between 12 and 1 o'clock, when the display was at its full brilliancy, the quiet streets of the city resting under this strange light, presented a beautiful as well as singular appearance.

Rudolf Wolf
(1816–1893)
Swiss astronomer, carried out historical reconstruction of solar activity back to the 17th century

David Hathaway, solar physics team lead at NASA's Marshall Space Flight Center in Huntsville, Alabama explains it this way:

> What Carrington saw was a white-light solar flare—a magnetic explosion on the sun. They give off the energy equivalent of about 10 million atomic bombs in the matter of an hour or two.
>
> The 1859 one was special, and it was noticed because it was a white light flare. It actually heated up the surface of the sun well enough to light up the sun.
>
> Now we know that solar flares happen frequently, especially during solar sunspot maximum. Most betray their existence by releasing x-rays (recorded by x-ray telescopes in space) and radio noise (recorded by radio telescopes in space and on Earth). In Carrington's day, however, there were no x-ray satellites or radio telescopes. No one knew flares existed until that September morning, when one super-flare produced enough light to rival the brightness of the sun itself.
>
> It's rare that one can actually see the brightening of the solar surface. It takes a lot of energy to heat up the surface of the sun!

Numerous extreme solar events have been reported since the 1859 Carrington event. In both 1921 and 1960 extreme radio disruption was experienced.

In 1972, a geomagnetic storm provoked by a solar flare knocked out long-distance telephone communication in Illinois. The main line from Chicago to San Francisco failed. Bell Labs researchers wanted to find out why, and their findings led them right back to 1859 and the auroral current.

A team led by Louis Lanzerotti went digging in the Bell Labs library for similar events and explanations. Along with field research, the history they uncovered became the core of a new approach to building more robust electrical systems in North America.

In 1989, an X19 class flare produced a storm that plunged six million people into darkness across the Canadian province of Quebec--and people were surprised when scientists later said a solar flare had probably caused it by overloading circuits.

In 2000, pager traffic in the U.S. was knocked out for a day, ap-

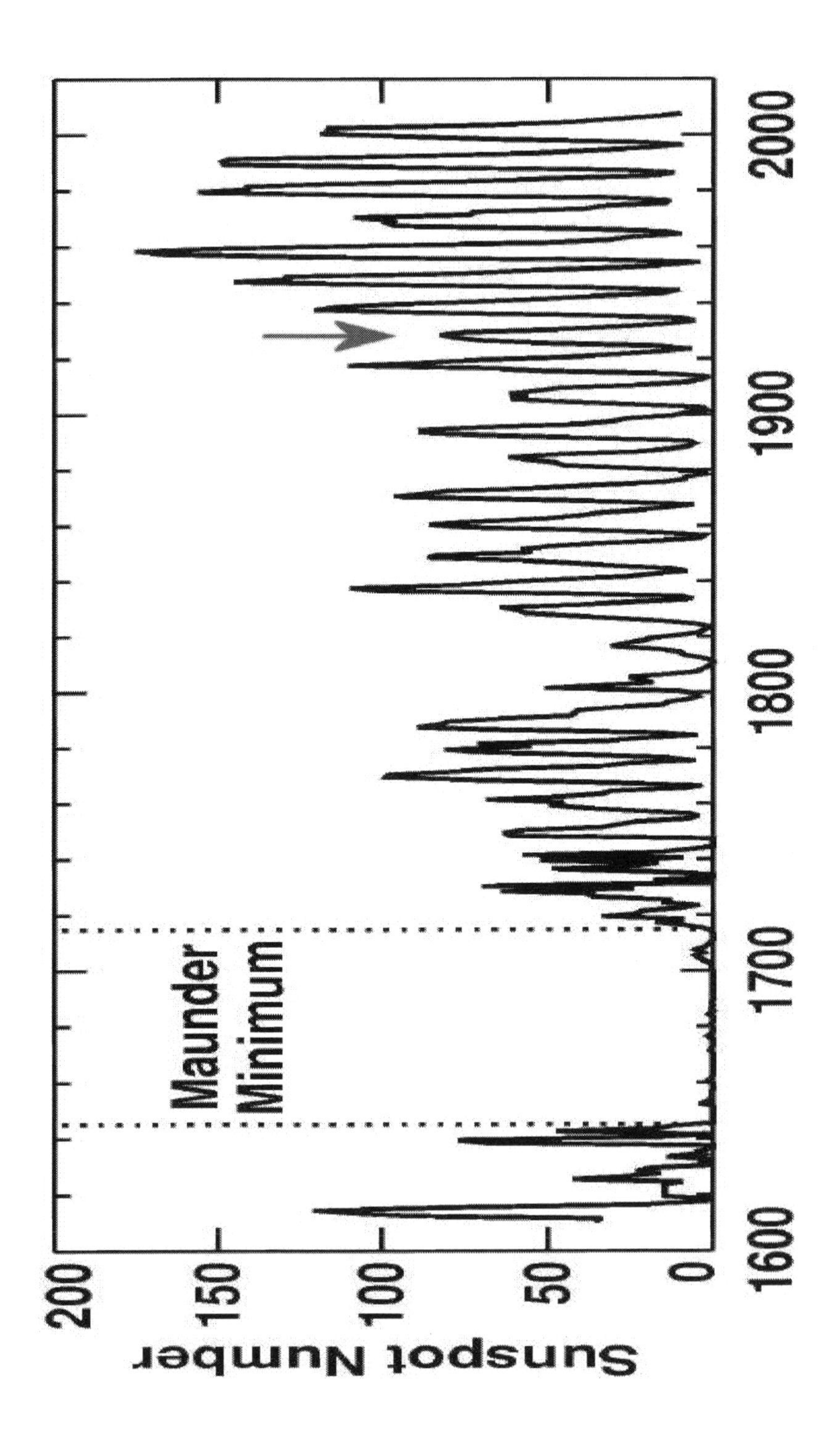
Sunspot Number
200
150
100
50
0
1600
1700
1800
1900
2000
Maunder
Minimum

parently because of a communications satellite that got fried by solar radiation in orbit.

The incidences of electronic equipment being disrupted or destroyed by solar flare activity are so common that it would be tedious to relate them all here. Many equipment failures remain speculative as to the cause of failure, and are not well known to the public, while equipment anomalies in military devices are not publicized at all (for good reason). On page 37 is a list of some of the most blatant occurrences in the last 163 years.

Scientists have been expecting these magnitude of events to become more frequent now (2011), as the sun is ramping up activity as part of its normal 11-year cycle. The solar maximum for this cycle is predicted to crest around 2013. Solar flare effects from increased solar activity have been on the rise for the last 100 years or so, and reliable research indicates that sunspot activity is directly correlated with warmer temperatures on earth.

Furthermore, based on research done on ice core samples from Greenland, a team of scientists headed by Sami Solanki at the Institute of Astronomy in Zurich, Switzerland, has re-constructed a history of solar flare effects and the sun's activity in the past. This research demonstrates that over the last century, the number of sunspots increased as the Earth's climate became gradually warmer. In addition to the above findings, these Swiss scientists have also determined that the sun's solar flare activity has been greater in the last 60 years than in the previous 1,000 years.

In 2005, Solanki stated: "Except possibly for a few brief peaks, the Sun is more active currently than at any time in the past 11,000 years." The physicist informed his colleagues that since 1940 the Sun has produced more sunspots, and also more flares and eruptions than in the past. Solanki published an earlier version of these findings in *Nature*.

From a *New Scientist* magazine article in November 2003:

> There have been more sunspots since the 1940s than for the past 1150 years (combined)." That is a 1825% increase. Sunspot numbers were derived from levels of a radioactive isotope found in ice cores taken from Greenland and Antarctica. Sunspots are the precursors of solar flares and coronal mass ejections and reflect the internal state of the sun.

The 1755--1766 cycle is considered the first cycle, and is called cycle number 1. So far, solar cycles have averaged 10.66 years in duration. Cycles as short as 9 years and as long as 14 years have been observed. Solar maximum in cycle 23 occurred in 1999-2000, with solar minimum in 2005. As of 2011, we are now in cycle 24, and are approaching the solar maximum, which has been predicted to peak sometime in 2013.

We all remember the Halloween solar storms, which took place between October 26th and November 4th 2003. For the first time in recorded history, two Saturn-sized sunspots appeared on the Sun's face at the same time. Both then proceeded to explode repeatedly with X-class flares. The storms climaxed on November 4th with the largest flare ever recorded, an X45. The CME's ejected by the Halloween storms were not Earth directed, but some ramifications were still felt here on Earth.

2005 was a particularly active year for the Sun. On January 17th, 2005 (during cycle 23's solar minimum) sunspot 720 produced an X-3 class solar flare. Three days later on January 20th, 720 spit out a huge X-7 class flare, which created a proton storm that reached Earth in a little over 30 minutes rather than the usual day or two. Scientists were shocked. CMEs usually travel at a rate of 1,000 to 2,000 kilometers per second.

"CMEs can account for most proton storms, but not the proton storm of January 20th," said Robert Lin, solar physicist at UC Berkeley.

Light from the Sun, traveling at around 300,000 kilometers per second, takes about eight minutes to reach the Earth, meaning that, in order for the protons pushed from sunspot 720 to have reached the Earth in thirty minutes, the particles must have been traveling at a quarter the speed of light, or about 75,000 kilometers per second. When anything travels at significant fractions of the speed of light, it is termed relativistic, referring to Einstein's fundamental relativity rule that matter cannot travel faster than the speed of light. Any particle traveling at the speed of light would achieve infinite mass. At even a fraction of the speed of light, the mass gets much heavier. So those protons, instead of being nearly weightless, would have sandblasted the Earth with considerable force.

Scientists were baffled, and are still speculating as to how the proton storm from sunspot 720 reached Earth so rapidly in 2005.

The year 2005 continued to be stormy, climaxing in September with one of the most turbulent weeks in recorded solar history. On September 7, sunspot 798, returning from the far side of the sun, unleashed a X17 class flare. The blast caused a blackout of many short-wave, CB, and ham radio transmissions on the day-side of the Earth, which at the time included most of the Western Hemisphere. Nine more X-class flares erupted out of the Sun over the next seven days; several spurred radiation storms that pelted the Earth.

On June 7th, 2011, one of the largest eruptions ever recorded spewed from the surface of the Sun. What shocked scientists was the unusual amount of material that lofted up, expanded, and fell back down over roughly half the surface area of the Sun. What also alarmed the SDO scientists was the fact that this particular sunspot was not unusually active prior to the eruption.

As mentioned earlier, on the morning of August 9th, 2011, at 0805 UT, sunspot 1263 produced an X7-class solar flare--so far the most powerful flare of new solar cycle 24. The brunt of the explosion was not Earth-directed. Even so, a minor solar proton storm enveloped our planet, and the flare also created waves of ionization in Earth's upper atmosphere.

Less than a month later on September 6th, 2011, active sunspot 1283 produced two major eruptions including an impulsive X2-class solar flare. The blasts hurled a pair of coronal mass ejections (CME's) toward Earth.

Two weeks later, on September 24th, sunspot 1302 crackled with strong flares, producing two X-class flares, followed by an M7-class eruption.

The activity continues. We must remember that from a long-term historical perspective, X-class solar flares are not common.

The "X" in X-class means"extreme".

Samuel Heinrich Schwabe
(1789–1875)
German astronomer, discovered the solar cycle through extended observations of sunspots

Major Historical Solar Events That Affected Electronics

1848 - Auroral 'earth currents' disrupt telegraph lines worldwide
1859 - Carrington event the largest CME event recorded to date
1860 - Coronal Mass Ejection first spotted during a total solar eclipse
1909 - Telegraph systems upset worldwide
1915 - Wireless outage in Northern Europe influences World War I
1919 - Whistlers heard for first time in World War I.
1940 - Easter Day shortwave disruption affects millions of phone calls
1941 - Major shortwave disruptions during World War II activities
1943 - Sunspots hamper radio transmission of Allied invasion of Italy
1947 - World-wide radio blackout. Airlines affected in Ireland
1950 - Early Korean War communications disrupted
1956 - Intense Cosmic ray blast and the British submarine incident.
1957 - World-wide radio fadeout lasts several hours on April 16th.
1959 - Shortwave blackout over North Atlantic on March 29th.
1972 - August 4 - Apollo 17 major flare.
1989 - March 8-13 Major solar flare, - Quebec Electrical Blackout
!994- - Jan. 20th Satellite failures
1997 - Jan. 11th 1997 Telstar 401 satellite fried
2000 - "Bastille day" storm effects the International Space Station
2001 - "Easter" storm X20 produced extensive satellite damage
2003 - November 4 - Major X28 flare ends major solar storm episode
2005 - Solar "minimum" year with multiple events--X17 class flare
2011 - Feb. 14th large CME erupts from the Sun

The shuttle Endeavor docked on the International Space Station.
Courtesy of NASA

Our Delicate Spacecraft

Flight system errors in spacecraft, commercial avionics and military avionics are very common. The majority of these errors have been attributed to single event upsets (SEUs), and many of those are said to have been caused by either cosmic radiation or energetic neutron particles emanating from the Sun. Roughly one in ten avionics errors are 'unconfirmed' which means that no obvious hardware or software problem could have caused them.

Billions of dollars of valuable satellite real-estate has already fallen victim to the effects of cosmic rays and solar storms. There are over 1000 operating satellites in space, worth an estimated $200 billion to replace. They account for nearly $225 billion in revenue for the international telecommunications industry every year.

NASA engineers have been aware of energetic particles in space ever since the launch of the *Explorer 1* satellite in 1958. James van Allen installed a simple Geiger counter device in the satellite which registered a surprising amount of particle activity. They found out that space was indeed radioactive.

> "The space environment is hostile. It is not a benign vacuum into which an object can be placed and be expected to operate exactly as it did on the ground." A. Vampola--physicist for the Aerospace Corporation

High-energy particles from solar flares or cosmic rays enter computer chips and change data or command bits from 0 to 1 or vice versa.

In low-Earth orbits, most of these events happen in the so-called

South Atlantic Anomaly (SAA) zone located over Brazil. This is where Van Allen radiation belt particles come closest to the Earth's surface.

> The following is from an article by Chung Yu-Liu in the IEEE AESS Systems magazine.
>
> It is well known that space radiation, containing energetic particles such as protons and ions, can cause anomalies in digital avionics onboard satellites, spacecraft, and aerial vehicles flying at high altitude. Semiconductor devices embedded in these applications become more sensitive to space radiation as the features shrink in size. One of the adverse effects of space radiation on avionics is a transient error known as single event upset (SEU). Given that it is caused by bit-flips in computer memory, SEU does not result in a damaged device. However, the SEU induced data error propagates through the run-time operational flight program, causing erroneous outputs from a flight-critical computer system.

The problems with SEU's is old news to the big players in the microchip industry, many of whom have been working to create radiation-hardened computer processors and components for many years. Companies and organizations like Intel, Boeing, Fairchild, RCA, NASA, Sandia National Laboratories, United Technologies Research Center, European, and Russian companies have been working on these issues basically ever since we started to send IC's into space.

As an example: Problems with SEU's have continually plagued NASAs TDRSS-1 satellite, which was launched in 1983. The satellite anomalies affected the spacecraft's Attitude Control System which, if left uncorrected, could lead to the satellite tumbling out of control.

Ground controllers have to constantly keep watch on the satellite's systems to make certain it keeps its antennas pointed in the right direction. This has become such an onerous task that one of the ground controllers, the late Don Vinson, once quipped, "If this (the repeated SEU's) keeps up, TDRS will have to be equipped with a joystick."

The circuitry in IC computer chips is becoming smaller every year, The smaller the circuitry becomes the more susceptible they are to the effects of solar flares and cosmic rays.

However, solar flare events can harm or effect much larger components, such as full sized transistors, which became evident in 1962, when the Telstar 1 satellite suddenly ceased to operate. From the data returned by the satellite, Bell Telephone Laboratory engineers on the ground tested a working twin to Telstar by subjecting it to artificial radiation sources, and were able to get it to fail in the same way. The problem was traced to a single transistor in the satellites command decoder. Excess charge had accumulated at one of the gates of the transistor, and the remedy was to simply turn off the satellite for a few seconds so the charge could dissipate. This, in fact, did work, and the satellite was brought back into operation in January, 1963.

Most manned spacecraft have about one quarter inch thick hulls, which provides a minimal amount of shielding for the astronauts and electrical components. On the surface of the Earth we are shielded by the planet's atmosphere, which provides shielding equivalent to several meters of solid aluminum.

The amount of shielding our atmosphere provides depends on where we are. Higher elevation sites are soaked in quite a bit more radiation than at sea level.

Scientists working at the Rutherford Appleton lab have shown that it is possible to generate a "portable magnetosphere", or magnet force field which would prevent ionized particles from reaching a space ship. ("Shields up Scotty !"). The plasma physicists have worked on the design which utilizes dipole magnetic fields instead of the typical heavy metal shielding. The current design however, requires a considerable amount of energy to generate the force field, which leaves it as yet unpractical.

Most manned activity involving the Space Shuttle and Space Stations occurs between 200 to 500 miles from the surface of the Earth. This very low orbital area is considered too close for most satellites because of the short 90 minute (average) orbital period. Communication satellites in that region would not be overhead long enough for ground interfaces to triangulate with efficiently.

Most of the newer satellites operate in what is known as *Low Earth Orbit* (LEO), which is the zone from 400 to 1500 miles from Earth. Pound for pound, the LEO zone is the most economical zone to place a satellite. Unfortunately, electronic devices that operate in LEO are more susceptible to magnetic storms than those operating in higher orbits.

The zone from 6000 to 12,000 miles from Earth is known as the *Mid Earth Orbit* (MEO). The GPS systems and some telephone satellites operate in MEO space, as the old *Telstar 1* once did. It costs more for companies to place satellites in MEO space than in LEO orbit.

Finally, at roughly 22,300 miles from the surface of our planet is the Clarke Belt of geostationary orbits (GEO). Even though it is much more expensive (by a factor of ten) to place equipment in GEO space, there are currently about 800 communications and military satellites still operating (and inoperative) in that zone.

Unexplained problems with spacecraft are common. Prior to 1998, Joe Allen and Daniel Wilkinson at NOAA's Space Environment Center kept a master file of reported satellite anomalies from commercial and military sources. Their library included over 9000 incidents. Within a years time, the information coming in to their database dried up. It appears to have been a change of guard; from the older scientific/military satellite builders to the nouveau slick communications start-ups. The old guard was in the habit of sharing data, while the new companies chose to be secretive. The novelty of satellite ownership had became passé.

Today communications companies are under pressure to build and launch high tech devices into orbit at the lowest possible cost.

This has resulted in a new attitude towards satellite manufacturing techniques that are quite distinct from the strategies used decades ago. Company technicians are pressured by sales departments to get new satellites into operation as quickly as possible. Competition between communications companies is fierce.

Seasoned satellite engineers are shocked at the ad hoc methods used to design and produce newer satellites. The older satellites may have been of more elementary designs, but they were usually built more robust.

Older satellites were built typically with mil-spec (military specifications) components, which represented designs of the highest quality. In most cases the component designs included considerable radiation tolerance.

Modern satellites are built with about 80% off-the-shelf components. Common components are cheap, available, and an irresistible lure for satellite manufacturers working within fixed or diminishing budgets. However, manufacturers willing to use mil-spec equivalent parts may find some parts unavailable, as demand has dried up, or newer tech mil-spec components have simply not yet been produced. Engineers can therefore find themselves in a catch-22 situation.

Satellite failure due to ordinary malfunctions are common. As much as 10% failure is considered acceptable in the industry. In contrast, a major priority has become keeping to the launch schedule.

Modern satellites may be built with more sophisticated electronic designs, but they have become like a lot of our 21st century electronic gear---semi-disposable.

There are many different ways that CME induced spacecraft dysfunction or outright failure can occur.

Space-born solar panels normally loose efficiency over time due to the continual bombardment of energetic particles. Engineers have learned to make the panels oversized to accommodate for degradation factors. The damage however, does not necessarily continue smoothly over time.

Collisions between high-speed protons ejected from the Sun and the atoms of silicon in the solar cells cause the atoms to violently shift position. This rapid movement of silicon atoms produce crystal defects that decrease the efficiency of the cells.

During a very energetic solar storm the solar cell degradation can happen very rapidly in a satellite's solar panels. During the well-documented solar storm in October, 1989, the *GOES-7* satellite suffered a five-year, 50% mission lifetime loss from that event.

High-energy particles can also do considerable internal damage to spacecraft as incoming protons collide with atoms in the satellite walls and produce sprays of secondary energetic electrons that can penetrate deep into the interior of the spacecraft.

In engineer speak, this is called *internal dielectric charging.* If this continues inside of a spacecraft, some electronic component will eventually break down and the circuit will produce a discharge. In a nutshell, you've then got a high-voltage current running through circuitry where it is not supposed to be.

As mentioned earlier, wild current flows can reverse a computer

Astronaut performing maintenance
Courtesy of NASA

memory position from "0" to "1" or vis versa. As any programmer knows, this is the basis of computer programming, and is the equivalent to changing the command from "yes" to "no". We need not elaborate on how that may effect the functionality of any electronic system.

These occurrences are called *Single Event Upsets* (SEU's), and come in two flavors: hard and soft.

A soft SEU only changes a binary valve stored in memory, and can be corrected by simply re-booting the device.

A hard SEU however, can do irreparable damage to microcircuitry inside the spacecraft.

Earlier generations of communications satellites used magnetometer sensors for positioning the spacecraft. The sensors detected the local magnetic field of the Earth and compared that to internal data to keep the satellite pointed correctly. Those satellites used what is known as *geosynchronous orientation systems.*

A solar storm can disturb the magnetic field of the Earth, and at times change the position of it's magnetic poles considerably. This can literally throw geosynchronous satellites into a tailspin.

In 1989, intense solar weather caused some spacecraft to experience uncontrolled tumbling, while others had to be manually operated from the ground to keep them from flipping over.

Modern spacecraft designs incorporate CCD sensor chips into their positioning systems.

CCD sensors are familiar to us because they are the most important element in common hand-held digital cameras. The CCD sensor sits exactly where the film in older cameras used to, and helps to create a digital image.

Spacecraft use CCD sensors by taking frequent images of the sky, and compare the location of detected stars to help re-orient the spacecraft. Energetic particles can impact the sensitive CCD camera elements, and produce false stars. The attitude systems in modern spacecraft are under particle bombardment constantly, which cause the devices to use unnecessary amounts of fuel as their pointing systems are frequently operating. During unusually large solar storms the sensitive CCD sensors can spit out incorrect data or be irreparably damaged.

What is in vogue today is for communications companies to place cheap satellites in Low Earth Orbit (LEO) and hope they function for awhile. Satellites operating in LEO are more susceptible to failure due to their proximity to the Earth's magnetic field. During a solar storm, high-energy particle intensity is greater in lower orbit zones.

When satellite failure occurs, the cause rarely can be well diagnosed. Dead satellites are almost never retrieved for inspection. Engineers can only speculate as to the cause of failure. Also, since there is no free-flow if information between industry, military, and scientists, the causes of equipment failure remains a frustrating thorn in the paw of space engineers. Competition has it's price.

Satellite insurers have therefore suffered, and have been quite vocal about it (insurance companies *really* do not like to loose money). The sparring between the underwriters and communications companies continues, while at the same time both endeavor to increase their profit margins.

Meanwhile, the public remains ignorant. We-the-people rarely hear about satellite failures. Neither the military, the underwriters, or the communications companies particularly want us to be aware of the full extent of it. The military have their own agendas, while the insurance and communications companies have stock holders to preen and fuss over. The communications companies have therefore jockeyed themselves into profitable, but very vulnerable positions.

The public was quite aware however, of the failure of the *Anik E2* satellite in 1994. The *Anik E1* and *Anik E2* satellites were a twin pair of 7000 pound GE Astro Space model 5000 units owned by Telestat Canada. In their day they were the most powerful satellites operating in North America, with virtually all of Canada's television traffic passing through them.

On January 20th 1994, NASA's *SAMPEX* satellite began to register some unusual solar activity. Within minutes the *GOES-4* and *GOES-5* weather satellites began to detect accumulating electrostatic charges on their surfaces. The *Intelstat-K* satellite then began to wobble, and experienced a brief outage of service. Like the *Anik E1* and *Anik E2,* the *Intelstat-K* was a GE 5000, except that it was built with radiation hardened components.

Two hours later the *Anik E1* began to roll over uncontrollably. When the *Anik E2* orbited into the firing line it also began to tumble. Both satellite's momentum wheel control systems failed, but the back-up momentum wheel on the *Anik E1* eventually started working seven hours later.

The *Anik E2* however, was another story. The back-up momentum wheel failed, and it took five months of hard work for Telestat engineers to gain control of the satellite. For the rest of it's life they had to manually control it by firing it's thrusters every one to two minutes. This worked only until the satellite ran out of propellent, which shortened it's life considerably. They were obviously reticent about the idea of scrapping a 300 million dollar piece of equipment. It was a sour victory though, because a few months later a critical diode in *Anik E1*'s solar panel shorted out, resulting in a permanent loss of half of that satellites power. Later, the Canadian engineers realized that this too had been a result of energetic solar particles. Ouch!

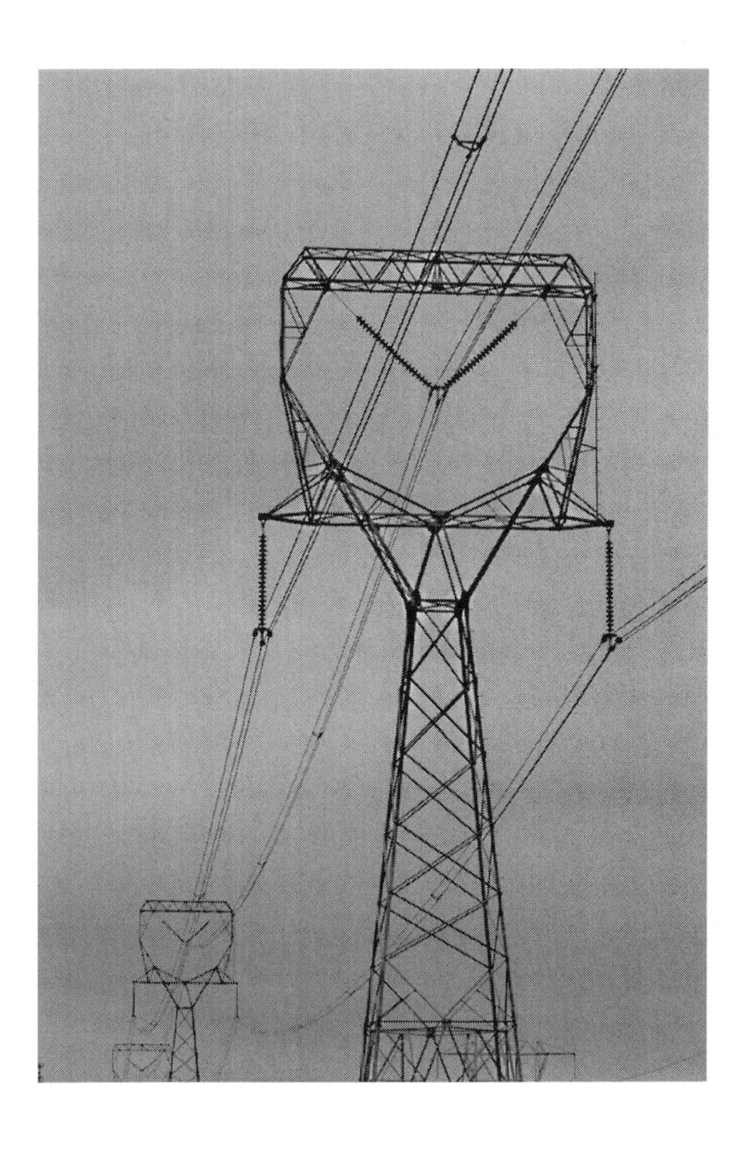

Our Fragile National Power Grid

Electricity is produced in hydroelectric, coal, nuclear, geothermal, solar, and wind generated power plants. The generated power proceeds to substations, which in turn feed high voltage transmission lines. The lines are supported by the tall metal towers we are all familiar with. There are over 150,000 miles of transmission lines in the U.S. The power travels down transmission lines, and is eventually fed into transformers on the ground. The transformers step down the voltage level from hundreds of thousands of volts to tens of thousands of volts. The current is then split up by using a device called a "bus", which is usually mounted near the transformer. From there the current is fed into local power lines, both overhead and underground.

The big transmission lines on their metal towers are perfect lightning rods for electrical surges from severe space weather to use as conduits. Power flows known as *geomagnetically induced currents* (GIC's) are induced by energetic particles known as *electrojets,* which swirl around inside the Earth's atmosphere.

Large power transformers in the US are particularly vulnerable to solar flare activity because many operate at high voltages--as high as 765,000 volts. Most of the high voltage transformers are single phase. Unfortunately, single phase transformers happen to be more susceptible to damage from geomagnetic activity than the typically smaller 3-phase models.

Trouble arises because the extra currents from solar storms are direct current (DC) flows, and the electricity transmission system is designed to handle alternating current (AC) flows. The extra DC flows saturate transformers, which start to overheat, causing their insulation to break down and their parts to experience accelerated aging. Above a certain temperature, a transformer will fail.

During the solar storm of 1989 the copper windings inside of a large transformer in the Salem nuclear plant in southern New Jersey were fused together. Fortunately, they located a spare transformer of the same type from a cancelled nuclear plant in Washington state.

Even then, it took 40 days to complete the removal and installation of the new unit. Typically these transformers are made-to-order type, and take from one to two years to manufacture.

Contrary to ordinary electrical engineering protocol, transformers in general tend to be susceptible to electrical surges *because* they are grounded. Many of the grids include transformers that have grounded neutrals. These transformer neutrals provide a path from the network to ground for these slowly varying electric fields to induce a current flow through the network phase wires and transformers.

During a CME event, transformers may be attacked from two fronts. One being from overhead electrojet activity, and the other comes from GIC subterranean currents that surge up through the ground rods connected to the transformers.

Large currents can be induced by electrojet activity through the highly conductive seawater in our oceans. These currents flow from the oceans into the land masses at different rates depending on the conductivity of specific areas. Much of the current will flow as deep as 400 kilometers, but some travel as deep as 700 kilometers under the surface of the Earth.

Electrojet footprints can be continent size and swirl around the poles in a latitudinal orientation. The orientation of the currents, both above and below the Earth is directional.

As a result of the natural tilt of the Earth's axis; every 24 hours the US and Canada find themselves right in the firing line of electrojet activity during a geomagnetic storm.

John Kappenman, with the Metatech corporation has made proposals to the U.S. Congress advocating the use of large 2.5 to 7 ohm resistors at the ground connection point of the transformers. This would protect the transformers from much of the currents surging up from the ground.

Unfortunately, the bill for the washing-machine sized resistors would be about $40,000 each, so electric companies are naturally hedging before making that kind of investment. There are over 5000 such transformers in the U.S.

Kappenman's testimony led the US House of Representatives to vote unanimously to fund his recommendations to protect the national power grid from dangerous solar storms that threaten to completely devastate our grid...and way of life.

> What follows is an excerpt from Kappenman's prepared statement for the House of Representatives Subcommittee on Environment, Technology, and Standards-Committee on Science on October 30, 2003:
>
> These currents (known as geomagnetically-induced currents-(GICs) are generally on the order of 10's to 100's of amperes during a geomagnetic storm. Though these quasi-DC currents are small compared to the normal AC current flows in the network, they have very large impacts upon the operation of transformers in the network.
>
> Under normal conditions, even the largest transformer requires only a few amperes of AC excitation current to energize its magnetic circuit, which provides the transformation from one operating voltage to another. GIC, when present, also acts as an excitation current for these magnetic circuits, therefore GIC levels of only 1 to 10 amperes can initiate magnetic core saturation in an exposed transformer.
>
> This transformer saturation from just a few amperes of GIC in modern transformers can cause increased and highly distorted AC current flows of as much as several hundred amperes leading to overloading and voltage regulation problems throughout the network.

Electrical engineers claim that if the Earth experienced another solar storm such as the one in 1921, it would burn out all 360 of the highest voltage transformers in the US.

Why didn't electrical systems go down more dramatically in 1921? The answer lies in the fact that in 1921 power plants and related systems were small, localized, and operated at much lower voltages.

Today, cities and regions share power via a massive connected system, which is efficient and flexible, but unfortunately quite vulnerable--particularly in the North American continent.

From Lawerence E. Joseph--writing for the New York Times:

> Storms don't have to be big to do damage. In March 1989 two smaller solar blasts shut down most of the grid in Quebec, leaving millions of customers without power for nine hours. Another storm, in 2003, caused a blackout in Sweden and fried 14 high-voltage transformers in South Africa. The South African experience was particularly telling — the storm was relatively weak, but by damaging transformers it put parts of the country off-line for months. That's because high-voltage transformers, which handle enormous amounts of electricity, are the most sensitive part of a grid; a strong electromagnetic pulse can easily fuse their copper wiring, damaging them beyond repair. Even worse, transformers are hard to replace. They weigh up to 100 tons, so they can't be easily moved from the factories in Europe and Asia where most of them are made; right now, there's already a three-year waiting list for new ones.

Surges and Spikes

These terms describe increases in the electrical power entering a system. A "surge" is a relatively slow build up of energy, and represents approximately a 3000 volt increase of a long duration. A "spike" occurs suddenly, and can produce 6000 volts or more above normal. In many areas of the U.S.A., it is not uncommon to record 1,000 volt increases daily over the normal 120 volts provided by the utilities.

Causes of voltage spikes can be attributed to many factors. Lightning strikes as much as several miles away can be conducted into an electrical panel as large voltage spikes. However, *under* voltage conditions ("brownouts") will also produce high voltage surges when power suddenly is restored. Typically, surge suppressors are only intended to protect equipment from lightning and power quality issues.

Ideally, alternating current (AC) switches from positive to negative in a smooth, even fashion. In reality, even in "normal" conditions electricity supplied to your home may occasionally have power quality problems, such as voltage surges, swells and sags and interruptions. These are called *transient* events and can be caused by weather, accidents, utility company equipment malfunctions, a neighbor's equipment, or appliances and tools at your own home. These variations had little effect in the vacuum-tube era of years past.

Today, modern electronic equipment is very sensitive to power quality problems, which can slowly damage or destroy semiconductor devices, such as microprocessors and dedicated-purpose electronic circuit chips.

The power supply in a computer, television, printer, or stereo (etc.) typically converts AC current into the DC current necessary to run the device, as well as determining how much voltage each component inside the device receives. A power supply prefers to have a reasonably consistent flow of AC current into it.

Surge protection devices have varying levels of performance and quality. High quality surge arresting equipment is recommended for our purposes. Any surge protector will eventually fail due to regular use, but the better ones hold up longer and under more duress. The circuitry in a surge protector regularly absorbs or sluffs off power transient events to protect the power supplies in the electronic devices it's hooked up to.

The performance of surge arresters is determined by two basic factors: the *clamping capability* (a term used to describe the level at which a unit begins to suppress and dissipate surges); and *response time* (how fast a unit can respond to the surge and begin "clamping" it off or readjusting the current to the voltage provided by the utility.)

One nanosecond is one billionth of a second, and is the measurement of electricity in light feet per second. To better comprehend such speed, consider one nanosecond as a wire 10 inches long. Many surge protection devices claim response times of five or more nanoseconds. This means that some voltage spikes would be over 4 feet past those surge arresters or suppressors and into equipment before they begin to react. In as much as the breakers in an electrical panel react in approximately 3 million nanoseconds, it should be apparent that electri-

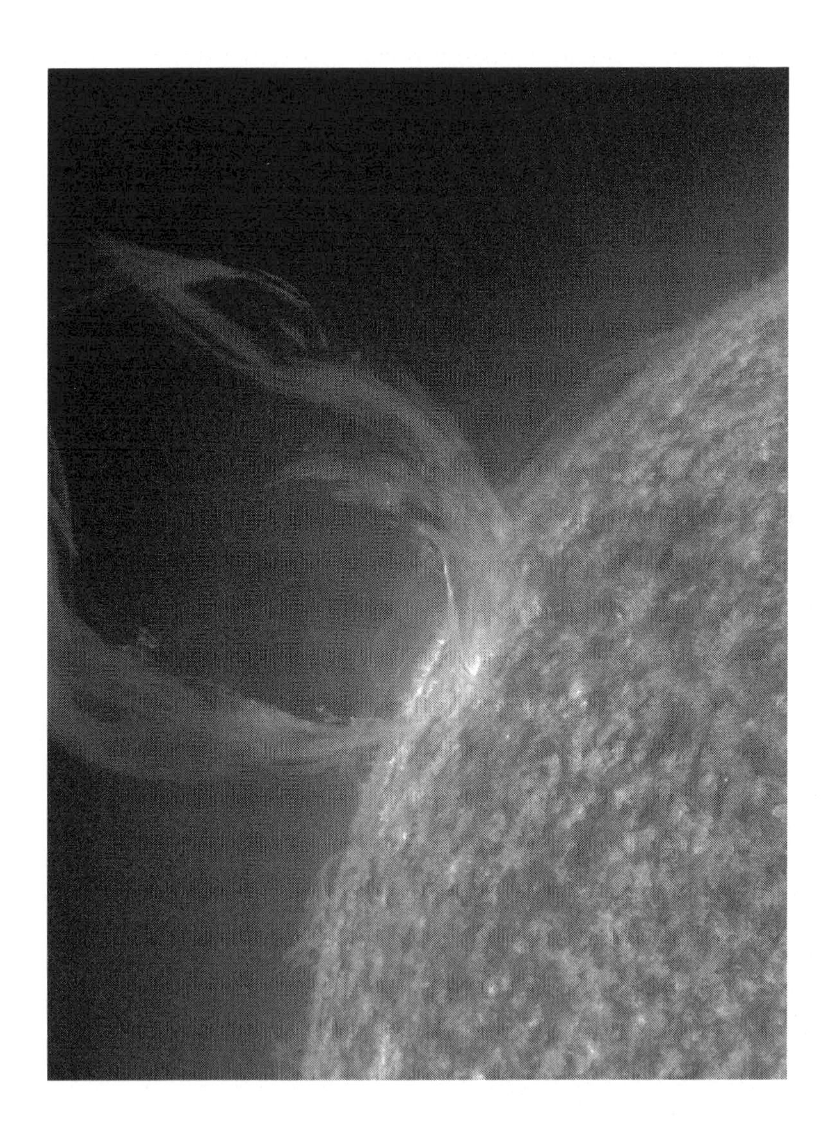

Solar Flare forming a loop
Courtesy of NASA

cal forces traveling at the speed of light can cause serious damage given these extra fractions of seconds.

Surge arrestors would probably prove ineffectual during a nuclear EMP burst, but may function adequately during a geomagnetic storm providing they have a response time faster than **one nanosecond**.

Once again however, no one really knows for sure what the Sun is going to throw at us--and at what intensity. Ideally equipment should be disconnected (unplugged) from power sources during intense solar events.

Some equipment is hard-wired to power sources, or has hard wired connection cables. Those cables should be shielded by running them inside grounded conduit.

Fixed, permanently installed generators should be grounded, and their output wires that feed into the electrical panel should be fully inside of grounded metal conduit that is mechanically attached to the panel.

Natural Gas Pipelines

Natural gas pipeline corrosion is a process that geomagnetic storms contribute to, given the right conditions and a lack of proper maintenance. Even though pipeline corrosion is not the biggest issue on our plate, it is certainly worth noting.

Modern pipeline engineers world-wide have been aware of these problems for years. According to studies done by the French national space agency CNES, pipelines designed to last fifty years can suffer wall erosion as much as ten percent in only fifteen years.

Many pipelines are partially protected from current flow effects by utilizing counter-currents of several amperes. The currents are oriented so that the pipeline has a net negative potential relative to ground. These systems only offer modest protection for the pipelines because auroral electrojet currents flow erratically, changing polarity constantly. During strong geomagnetic storms, currents as high as 1000 amperes have been monitored.

The Alaskan pipeline is particularly vulnerable to GIC's due to it's latitude and it's north/south orientation.

LG
SEND
2ABC
3DEF
4GHI
5JKL
6MNO
7PQRS
8TUV
9WXYZ

Our Disposable Personal Electronic Gear

"The sun is waking up from a deep slumber, and in the next few years we expect to see much higher levels of solar activity. At the same time, our technological society has developed an unprecedented sensitivity to solar storms."

Richard Fisher, head of NASA's Heliophysics Division

We can loosely divide the history of electronic circuitry into three categories: Vacuum tube gear (1858-1970), transistor equipment (1948-1980>), and integrated circuits (micro-chips) (1958>).

In days of yore, electronic equipment was powered by vacuum tubes. Those old radios, record players, and televisions tended to be fussy, but were reliable in spite of their quirky ways.

The circuitry was simple and repairable. I recall trundling off on my bicycle to the electronics store with handfuls of tubes from our "stereo" or black & white television "set". Once there, I would dutifully check out the function of each tube using the tube testing machines provided by the store. If a tube seemed faulty, I would cross-check between machines, and if necessary, purchase a RCA 6NB6 ,a Sylvania 12AX7, or whatever tube I deemed needed replacing.

Vacuum tubes were expensive, as was the related equipment. Dollar for dollar, a 13" black & white television in 1962 cost the same as a 13" color TV 40 years later.

Inventors began to work on tube technology in the mid 19th century, but it wasn't until 1904 that the first practical electron valve tube was invented by John Ambrose Fleming. New tube designs continued to be created and refined until about 1960, when the focus had already shifted to transistor technology.

Vacuum tubes are still used in various industries. Transmitter-type tubes are still favored by some because they are able to survive lightning strikes better than transistor transmitters do. That is due to the fact that vacuum tubes can handle voltage surges better than transistors.

One of the most popular uses of vacuum tubes today is a sector that revels in delight with their controlled use of "voltage surges"--electric guitarists. Tube-type electric guitar amplifiers may be considered to be better sounding than transistor amps by guitar aficionado's, but it is the durability of a vacuum tube under stress that we are concerned with here.

The musicians' use of controlled tube stress is illustrated by the well known technique used by Jimi Hendrix on stage. Jimi would turn his tube-type amplifiers (plural) all the way up, then control the volume from his guitar. He wanted the tubes to be running as hot as possible to achieve the effect he wanted, which is known as "natural tube distortion". At the same time, when he would strike a note or chord aggressively, it would create a large voltage surge in the vacuum tubes and the rest of the circuitry.

Transistor guitar amplifiers do not tolerate that kind of abuse well. If driven too hard, a transistor amplifier will either "clip", or transistor failure can occur.

Inside of a vacuum tube, electrons flow between electrodes or metallic plates that are separated by short distances. Like in a light bulb, the flow of energy occurs inside the glass of the tube in a vacuum. When a surge occurs, the tube tolerates it, but energetic particles behave wildly, the whole system heats up, rapidly becoming less efficient.

Vacuum tubes are still used by various militaries in devices designed to withstand power surges originating from atomic explosions. Years ago when NATO captured one of the "modern" USSR fighters, everyone laughed--at first--about the seemingly antiquated tube type technology until they realized that the Russian fighter would still work after an EMP weapon attack and ours would be destroyed and useless on the ground. It's interesting to note that some of the Soviet Union's most sophisticated fighters of the cold war era were fitted with electronic systems that contained no solid state (transistor) devices at all, but instead used many, many miniature vacuum tubes fastened to circuit cards.

This discovery resulted in a lot of changes to our own military's equipment designs.

It should be noted that the earliest use of SS diodes, as signal detectors, began in the 1940s, so even some "all tube" vintage circuits contain SS diodes. These diodes have very high current ratings therefore would probably tolerate current levels induced by a GIC, but perhaps not from a nuclear EMP.

A transistor is a device composed of semi-conductor materials (mostly silicon) that can both conduct and insulate. The transistor was the first device designed to act as both a transmitter--converting sound waves into electronic waves, and as a resistor--controlling electronic current.

Transistors were invented in 1948 at the Bell Telephone Laboratories in Murray Hill, New Jersey by John Bardeen, William Shockley, and Walter Brattain. Later, in 1956, the team received the Nobel Prize in Physics for the invention of the transistor.

Transistors are small, efficient, light weight, and use much less power to operate than vacuum tubes. The large power consumption necessary to drive vacuum tubes requires heavy input and output power transformers. A 40 watt vacuum tube audio power amplifier can weigh as much as 50 pounds, while a comparable transistor unit may weigh only 75% less.

In 1952, the Sonotone hearing aid appeared on the market, making it the first use of the transistor in a commercial product. In 1954, the first transistor radio, the Regency TR1 was manufactured.

Compared to small tube-type radio's those early "transistor radios" were harsh and obnoxious sounding, which was a problem with transistor audio equipment until the 1970's. Even today, some audiophiles prefer the pure, smooth sound of vacuum tube stereo gear.

The transistor however, became the amplifier of the modern era, and has grown smaller incrementally with each decade.

In the vacuum tube era, most of the wiring inside of electronic equipment was run in what is known today as "point to point" wiring. The tube sockets and other components had lugs to which actual wires were connected, which formed the circuits. Visually, common equipment wiring was messy, with what seemed like wires running every

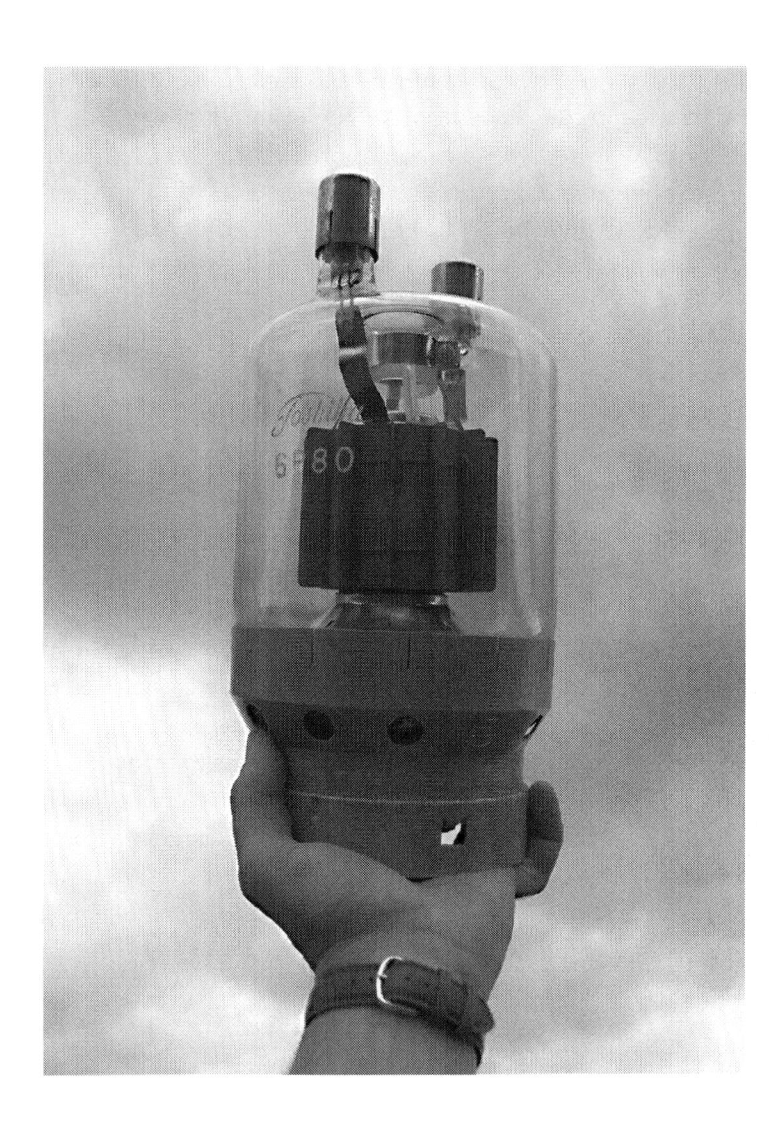

Large vacuum tube circa 1975
Tubes such as this one are still used in some transmitter applications

which way. However, the wires running through the air benefited from positive heat dissipation.

Printed circuits did not become commonplace in consumer electronics until the mid-1950's. Unlike the transistor, printed circuit boards may not have warranted Nobel prizes, but the invention certainly revolutionized electronic equipment design.

We are all visually familiar with printed circuit boards; those little colored plastic wafers with orderly embedded flat shiny copper wires connecting the capacitors, resistors, transistors, and diodes together.

Between 1936 and 1947 John Sargrove's Electronic Circuit Making Equipment (ECME) sprayed metal onto a Bakelite plastic board. The ECME could produce 3 radios per minute.

However, the invention of the printed circuit board has been attributed to the Austrian engineer Paul Eisler who, while working in England, made one in 1936 as part of a radio set.

Printed circuit boards are durable, and an efficient use of space. Meters of copper wire can be imprinted into the board in rows. However, during a powerful geomagnetic storm, the little flat copper wires can act as a long antenna, and provide a conduit for electricity to travel down and potentially burn out the parts connected to them.

The use of smaller transistors and printed circuits eventually evolved into much tinier formats known to us as integrated circuits (IC's).

Without knowing each other, through two independent paths in 1958, two engineers invented, almost at the same time, the Integrated Circuit (IC). The invention of Jack Kilby (Texas Instruments) and Robert Noyce (Fairchild Semiconductors), also known as "the chip", has been recognized as one of the most important innovations and significant achievements in the history of humankind.

The revolutionary idea Kilby and Noyce came up with was to imprint components (like transistors) right into the circuit boards using semi-conductive materials. The circuit boards could then be laminated one on top of another with an insulator in between. This development allowed for the creating of micro-sized circuitry, which has become the basis of modern electronic equipment. Today, a little 10mm microchip can have as much circuitry in it as an entire old table-top radio from the 1960's.

Electronic Devices Susceptible To Geomagnetic Storms

Automobiles
Aircraft
Power Boats
Semi trucks & Busses
Trains
Security systems
Electronic medical equipment
Pacemakers
GPS systems
Telephones and land line systems
Cellular phones and cell tower systems
Surge protectors
Batteries
Computers & peripherals
Broadcasting equipment
Televisions/ DVD and VHS players
Radios
Stereo/CD players/Tuners
Fax machines
Copy machines
Coffee makers
Refrigerators
Ranges
Dishwashers
Dryers & Washing machines
Power tools
Calculators
Quartz Watches

Virtually all of our modern consumer electronic gear is built with IC circuitry. The use of IC's have allowed us to have personal computers. Computers heralded the age of the internet, which, along with GPS systems and cellular telephones have opened up lines of communication we had not dreamed of a mere 50 years ago.

As we are all aware, we have become reliant on our electronic gear to an alarming degree, but understandably so. That little cell phone provides the owner with a remarkable amount of freedom. Without IC circuitry, we would be carrying our cell phones around in heavy backpacks, and they would not have nearly the variety of functions they do today.

Why is electronic gear that is built with IC circuitry particularly vulnerable to geomagnetic storms?

When any piece of electronic equipment is operating, it naturally produces heat. All of the components heat up, including the wires. That is due to the fact that electronic circuitry as we know it is not 100% efficient. If it was 100% efficient there would be no heat generated. Transistors in particular produce a substantial amount of heat. Transistors may operate much more efficiently than vacuum tubes, but they tend to get so hot that large transistors are often mounted on finned aluminum heat sinks or are externally cooled by fans. That is why dust is the enemy of all electronic equipment. Dust provides components with unwanted insulation, causing them to heat up to unnatural degrees.

I have personally seen an audio amplifier (being driven to its extreme limits) burst into flames. The transistors probably got so hot that they ignited the dust that coated the parts.

Micro chips are particularly vulnerable to heat because all of the components and wiring are crammed together into a tiny space. IC's are known to loose efficiency rapidly as they heat up. Any piece of modern electrical equipment left operating in a window sill in full sunlight is a candidate for the graveyard.

Heat is a perpetual problem with IC's, and could contribute to component failure during a GIC (geomagnetically induced currents), but it is those maverick currents roaming through the circuits that will destroy the equipment.

All solid state (transistor) circuits are vulnerable to GIC's but the tiny IC'c are by far the most sensitive. Anyone who has added RAM to their computer knows that just a little static electricity generated from a sloppy installation can burn out the circuit on the card. Most circuits are designed to perform one function, and have a certain amount of tolerance depending on the durability of the system. Tolerances are typically narrow, and currents running in the opposite direction will easily fry delicate components.

Communications and GPS Navigation Systems Vulnerability

The effects of solar storms on radio communications and broadcasts have been obvious since the early days of radio. Prior to the first observations of encounters with radio waves and the Sun, Guglielmo Marconi claimed his "wireless" communication systems were immune to solar activity. He was embarrassingly incorrect. Marconi was promoting wireless communications in favor of land line telephone systems. In spite of the visionary idea, land lines became the norm and have remained the most stable form of communications to date. We can only imagine Marconi's reaction to the prevalence of cellular telephone usage in the 21st century.

We need not elaborate on the vulnerability of cellular telephones to geomagnetic storms. Even during daily use, every cell phone user knows how fussy reception can be in different areas. It would not take much to upset that apple cart, but the real danger is to the handsets themselves, not to mention the transmitters working at the heart of the systems.

Much has been written concerning the variability of the accuracy of GPS navigation systems due to geomagnetic activity. An even slight alteration in signals from GPS satellites can result in inaccuracies of several meters or more on the ground.

Commercial ship navigators are familiar with anomalies, and regularly allow for them. However, at times they must navigate within tight parameters. The Exxon Valdez incident is an example of a navigation error probably caused by GPS inaccuracy. There *was* a significant solar event immediately prior to the wreck. The navigation error was less than one-third the length of the ship. Unfortunately for the ship's captain, the powers-that-be had a definite need to blame *someone.* As previously mentioned, underwriters and insurers loose sleep over, or have denial issues concerning "acts of nature".

Military and commercial airline pilots and navigators are also fully aware of navigational anomalies associated with geomagnetic storms.

GPS satellites operate in orbit 22,000 km above the Earths surface. Signals received by aircraft and ships from GPS satellites must pass through the electrically charged ionosphere.

Ionospheric Variability

On the sunlit side of the Earth, when solar radiation strikes the electrically neutral parts of the ionosphere, electrons are dislodged from atoms to produce the ionospheric plasma.

From about 500-1000 km above the surface of the Earth is the region called the ionosphere, which gets it's name from the highly ionized conditions that persist there. The ionosphere reacts dramatically to the intense x-ray and ultraviolet radiation released from the Sun during a powerful solar event. The continual blasts of particles and energy from the Sun strike the ionosphere so strongly that electrons are actually stripped away from their nuclei. The electrons and nuclei run around freely in a plasma state, which turns the upper atmosphere into an electrical conductor. This high conductivity factor makes the ionosphere an important, active part of the Earth's atmosphere, even though it makes up less than one percent of the mass of the atmosphere above 100 km.

The conductive character of the ionosphere has been used for over a century due to its influence on radio waves, which bounce off of it and shoot back to the Earth's surface.

The most important area in the ionosphere for communications and navigation systems is called the F2 layer, which is about 500 km from the surface, and is where electron concentrations reach their highest values.

Variability of signals that must pass through the electrically charged F2 layer is well known to GPS engineers, and is a more acute problem during geomagnetic storms.

Astronaut performing maintenance
Courtesy of NASA

Our Sensitive Bodies

Radiation. The word conjures up images of forboding, unseen dangers, and old dusty, rusting radioactivity signs hanging askew--blowing in the relentless desert wind. Our image is of Hollywood-inspired relics from the cold war era.

In reality, we are all bombarded by different forms of radiation every day. Our planet's atmosphere may buffer it, but still our bodies are constantly bathed in various types of atomic particles. Human-kind has evolved in this environment, causing us to have genetically developed repair mechanisms tailored to deal with modest levels of radiation.

How our bodies deal with radiation varies depending on the type and intensity. At lower levels our cells experience DNA damage but are able to detect and repair the damage. As the intensity increases our bodies suffer more, and may not be able to repair the damage.

The term "radiation" is used commonly to describe matter - in the form of electrons, protons and the ions of various atoms, which travel much slower than the speed of light.

Radiation research has been extensive. Medical studies concerned with the effects of radiation exposure to the human body could fill entire libraries.

After World War II, "sunbathing" became popular, and still is for the foolish few who dogmatically endeavor to improve their "tan". The rest of us are careful. The era of the solar worshippers is over.

The Sun is not the only source of radiation we are subjected to daily. Cosmic rays arrive from deep space after travelling for billions of miles at near-light speed. These alien invaders arrive here in many forms and species, and can be surprisingly damaging to electrical equipment in space, and the human body on Earth.

When a star finally grows old and dies it can explode, forming a supernova. These interstellar detonations fill space with high energy particles that blast out in all directions.

Our solar system's magnetic field serves as a partial shield for cosmic rays, deflecting the less energetic particles. Our next line of defense is the magnetic fields of the planets themselves, who, like watchdogs barking at intruders ward off even more of the atomic invaders. Eventually the most determined cosmic rays make it all the way into the Earth's atmosphere where they collide with nitrogen and oxygen atoms to produce secondary showers of energetic particles. We are bathed in a steady rain of cosmic droplets day in and day out. Depending on where we live, we absorb from 35 to 130 millirems per year.

The influx of cosmic rays varies depending on the 11 year solar cycles. Cosmic ray particles volumes entering the Earths atmosphere rises and falls exactly out of step with the 11 year cycles. During solar maximum periods, when the Earth tends to be the recipient of more solar radiation, cosmic ray intensity is less. When the Sun is particularly active it's magnetic field is stronger, which in turn wards off more interstellar particles.

Recent studies by Dr. Denis Kucik at Brookhaven lab in New York suggest that high energy iron ions from cosmic rays could pose unique problems for shielding deep-space spacecraft. Iron ions have the ability to smash through shielding, generating sprays of secondary radiation.

Earlier studies indicate that iron ions from cosmic rays may encourage hardening of the artery walls encouraging Atherosclerosis.

As mentioned earlier, on the surface of the Earth we are shielded by the planet's atmosphere, which provides a shielding effect equivalent to several meters of solid aluminum.

We subject ourselves to much higher levels of radiation when exposed at high elevation sites, while normal atmospheric shielding protects us more at sea level. Mountaineers, skiers, and mountain folk should be aware of current radiation levels. As mentioned earlier, airline workers are particularly susceptible to health issues associated with radiation.

Some airline companies have protocols as to how many hours workers are allowed to spend in the air per month.

Long term radiation exposure to commercial airline pilots and flight attendants is considerable, and well documented. Once again, the amount of radiation encountered when traveling in a commercial airliner varies depending on the route of the flight. Flights over polar regions bathe travellers in more radiation than in equatorial areas.

The amount of radiation absorbed by travelers on a routine flight is about the same dose of radiation one receives from having a full body x-ray in a hospital.

The *Thermosphere* is a layer of Earth's atmosphere that ranges in altitude from 90 km to 600+ km. The thermosphere intercepts extreme ultraviolet (EUV) photons from the Sun before they can reach the ground. When solar activity is high, solar EUV warms the thermosphere, at times reaching temperatures as high as 1400K.

The thermosphere normally contracts and expands in sync with solar minimum and solar maximum periods. During solar maximum it heats up and expands, and during solar minimum it cools and contracts.

During the recent solar minimum period in 2008 the thermosphere contracted much more than expected. "This is the biggest contraction of the thermosphere in at least 43 years," says John Emmert of the Naval Research Lab, lead author of a paper announcing the finding in the June 19th issue of the Geophysical Research Letters (GRL).

Emmert used a clever technique. Because satellites feel aerodynamic drag when they move through the thermosphere, it is possible to monitor conditions there by watching satellites decay. He analyzed the decay rates of more than 5000 satellites ranging in altitude between 200 and 600 km from 1967 to 2010. This provided a unique space-time sampling of thermospheric density, temperature, and pressure covering almost the entire Space Age. In this way he discovered that the thermospheric collapse of 2008 was not only bigger than any previous collapse, but also bigger than the Sun alone could explain. "Something is going on that we do not understand," says Emmert.

The Earth's magnetic fields harbor energetic particles in a complex whirlwind of activity in a region known as the *Plasmasphere,* which is further away from the planet than the highest orbits of satellites.

Within the plasmasphere, high energy protons and electrons bounce back and forth along their northern and southern loops. Meanwhile, electrons flow eastward while protons move westward in two powerful intermingled "ring currents".

Contrary to earlier models, we now know that much of the energetic particles in the plasmasphere come from Earth, not from the solar wind, as was previously thought. Energetic particles gush out of the ionosphere in great polar fountains and are deposited in the plasmasphere. This process is not well understood, but somehow particles are accelerated to very high energies and can be held at temperatures of thousands of degrees.

The plasmasphere is therefore, a very dangerous environment for humans to be in. An un-shielded astronaut lingering in the plasmasphere would be zapped with 1000 rem of radiation per hour.

Fortunately, most of us will never venture that far away from the surface of our planet. The mysterious mechanics of the plasmasphere is but one example of the fascinating geomagnetic workings of planet Earth.

On January 20th, 2005, a giant sunspot named "NOAA 720" exploded. The blast sparked an X-class solar flare, and hurled a billion-ton cloud of electrified gas (CME) into space. Solar protons accelerated to nearly light speed by the explosion and reached the Earth-Moon system minutes after the flare--the beginning of a days-long proton storm.

The Jan. 20th proton storm was by some measures the biggest since 1989. It was particularly rich in high-speed protons packing more than 100 million electron volts (100 MeV) of energy. Such protons can burrow through 11 centimeters of water. A thin-skinned spacesuit would have offered little resistance.

The radiation risk posed by solar flares and their bad-boy CMEs is one of the major concerns in discussions of manned missions to Mars or to the moon. More effective physical or magnetic shielding would be required to protect the astronauts. Originally it was thought that astronauts would have two hours of time to get into shelter, but based on the 2005 event, they may have as little as 15 minutes to do so.

Solar Flares Affect Stock Markets?

In 2003, Anna Krivelyova and Cesare Robotti , working for the Federal Reserve Bank of Atlanta wrote an extensive report about the effects of solar flare activity on the world-wide stock markets.

The technical paper is titled: *Playing the Field: Geomagnetic storms and International Stock Markets.*

Using data from 1932 to 2003 from many countries and stock indices, they used various different models to compare geomagnetic storms to major and minor moves in the markets.

In a nutshell, they found that markets tended to go up during solar minimum periods. This occurs most dramatically with the large caps like the S&P 500 and the Dow Jones industrials. Inversely they found that markets tended to drop during solar maximum (particularly small caps like the Russell 500).

More specifically, at least in the short term, the data suggested that markets usually go down during and a few days after large geomagnetic events.

We must point out that these folks are serious, as is their employer. One can make a reasonable uneducated guess as to which deity the Federal Reserve Bank of Atlanta worships--the almighty dollar.

In 2003 there was a lot of work being done to program algorithmic computerized trading systems designed to trade on the Forex (currencies), commodities, and stock markets. Today those systems have become a reality. It has been rumored that Goldman Sachs has a market-trading "robot" that wins *every* time.

Why do solar flares affect the stock markets? Krivelyova and Robotti's research suggests that it has to do with the human body's physiological reaction to solar storms--causing individuals to make specific types of decisions. It seems that stock traders make poor choices during solar storms.

From the report mentioned above: (Krivelyova and Robotti)

> For example, the average number of hospitalized patients with mental and cardiovascular diseases during geomagnetic storms *increases* approximately two times compared with quiet periods. The frequency of occurrence of myocardial infarction, angina pectoris, violation of car-

dial rhythm, acute violation of brain blood circulation doubles with magnetically quiet periods". The report also states that "At least 75% of geomagnetic storms caused *the* increase in hospitalization of patients with the above mentioned diseases by 30%-80% at average.
Solar flares affect the Central Nervous System (stomach lining), all brain activity(including equilibrium), along with human behavior and all psycho-physiological (mental-emotional- physical) response. Solar flares can cause us to be nervous, anxious, worried, jittery, dizzy, shaky, irritable, lethargic, exhausted, forgetful, have heart palpitations, feel nauseous and queasy, and last but not least, to have prolonged head pressure and headaches. There are reports of visual disturbances, inner ear issues, buzzing in ears, throat and thyroid issues, cold feet, and even a weird symptom of having tongue dryness.

While no one fully understands exactly why or how the brain responds as it does to electrical currents and magnetic waves, intriguing new research is offering some possible explanations.

A recent study published in the science journal *Nature*, indicates a direct connection between the Sun's generating of charged particles such as solar flares, CME's, and coronal holes which on occasion open wide gaps of highly charged Sun particles.

Studies have shown that the right temporo-parietal junction (TPJ) lights up with activity when we engage in moral judgments like evaluating the intentions of another person. This indicates that the region is important to making moral decisions. But while we like to think we're very consistent in our morality, an MIT team showed that an electromagnetic field applied to the scalp impairs our ability to evaluate the intentions of others, leaving us with little by which to hand down a moral judgment.

It's a completely different approach because of the way the magnetism is applied to the brain. But it's an example of burgeoning new research on an old idea; that the brain is an electromagnetic organ and that brain disorders might result from disarray in magnetic function. The idea has huge appeal to psychiatrists and patients alike, since for many people the side effects of psychiatric drugs are almost as difficult to manage as the disease itself. Almost 30 percent of the nearly 18.8 million people who suffer from depression do not respond to any of the antidepressants available now. People with other severe mental disorders--schizophrenia, obsessive-compulsive disorder--might benefit as well. And while no one fully understands exactly why or how the brain responds as it does to electrical currents and magnetic waves, intriguing new research is offering some possible explanations.

Robert J. Gegear, lead researcher from the Department of Neurobiology, University of Massachusetts Medical School reports:

> Our study suggests humans may be genetically pre-disposed to the influence by geomagnetic flux as it relates to the Earth's magnetic field and charged particles such as solar flares, coronal mass ejections, gamma rays, and galactic cosmic rays.

Another new study shows consistent evidence of an influence of geomagnetic fields on the light sensitivity of the human visual system. It has been proposed recently that light-sensitive magnetic responses are not only used for directional information, but may also aid as a 'visual' barometer by providing a spherical coordinate system for integrating spacial awareness.

The dormant genes residing within all of us just ready to be tapped are called *Cryptochromes*. They are a class of blue light-sensitive flavoproteins found in plants and animals. Cryptochromes are involved in the circadian (24 hour cycle) rhythms of daily life.

Marrow-containing bones, such as those found in the skull, shoulders, spine, sternum, and thighs, are the portions of the body most vulnerable to radiation. Solar protons damage the blood-producing cells living in the marrow, depleting the body's fresh blood supply.

The effects of radiation on the human body have been very well researched, with new information surfacing regularly. We all know the basics. Depending on ones skin type, regular moderate doses of normal sunlight can be beneficial to our bodies, while excessive amounts of Sun exposure can be considered an overdose, and harmful to us in many ways.

At the risk of being redundant--stay out of the sun during a powerful solar event! Consider it a good opportunity to spend some time relaxing in the shade of a stone, concrete, or brick wall; on the shade side of a hill, or in your Faraday-type shielded environment.

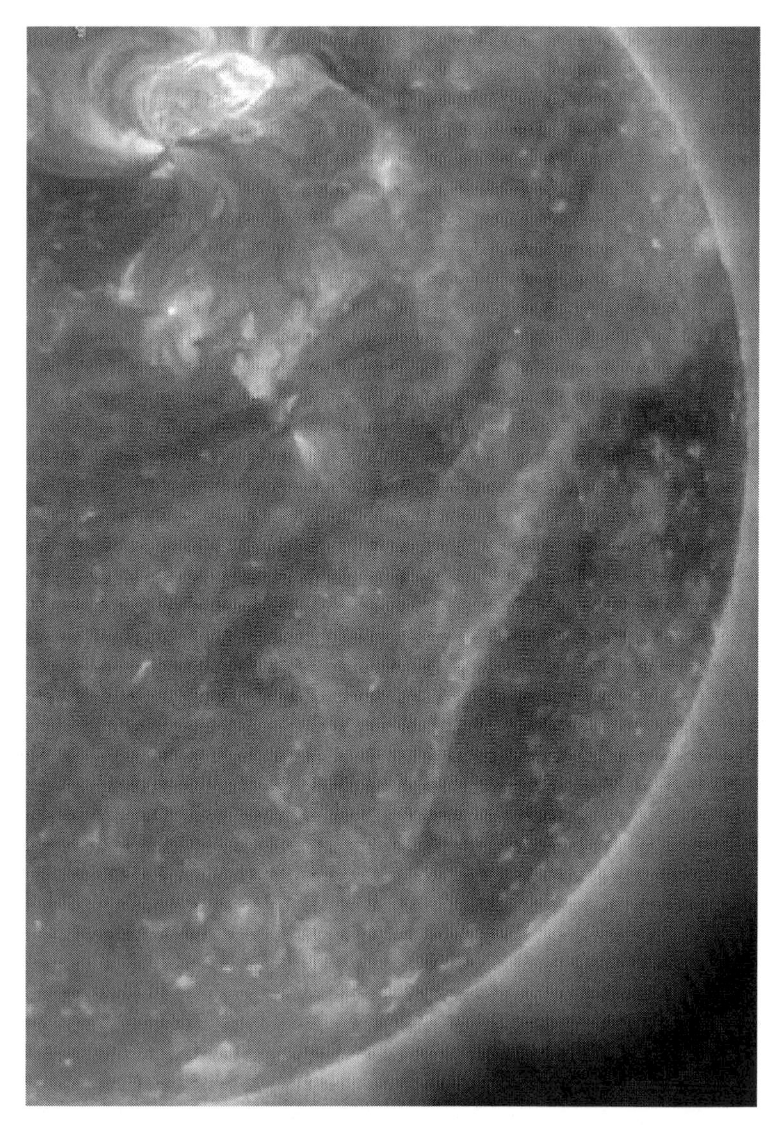

On August 1st, 2010, almost the entire earth-facing side of the sun erupted in a tumult of activity. There was a C3-class solar flare, a solar tsunami, multiple filaments of magnetism lifting off the stellar surface, large-scale shaking of the solar corona, radio bursts, a coronal mass ejection and more.

Courtesy of NASA

Can Solar Flares Trigger Earthquakes?

The world watched in horror as multiple tsunami's rolled over the east coast of Japan on in the afternoon of March 11th 2011. As the events unfolded, many quakes such as several over 5.0 Richter off the coast of California that normally would have received attention were ignored in favor of the devastating Japanese event.

Immediately after the 9.0 quake off the coast of Japan, Mt. Sakurajima erupted, and at the same time two Russian Volcanos--Kizimen and Sheveluch went off as well. A few hours later Mt. Karangetang in Indonesia erupted.

24 hours prior to the 9.0 quake in Japan, 25 people were killed in an earthquake near the China/Myanmar border in Yunann Province.

Examining the world-wide seismic activity from the USGS site was particularly revealing.

This extremely seismically active period actually began on March 9th with a "foreshock" quake measuring 7.2 on the Richter scale off the coast of Honshu, Japan. What followed were over 700 quakes in the same area over a period of one week. These earthquakes averaged about 5.5 on the Richter scale and most occurred offshore at a depth of about 25 km slightly inland of the Japan Trench. On the 15th there were still quakes registering as high as 6.2. The 9.0 quake on March 11th was the most dramatic, and well documented by the news media.

Gerald Fryer, geophysicist for the Pacific Tsunami Warning Center claimed that the 3/11/11 quake was actually a magnitude of 9.1. The new magnitude was adjusted based on the impact of the quake throughout the Pacific, he said. "It fits all measurements, including in Hawaii," Fryer said.

The U.S. Geological Survey estimate of the quake's magnitude is still 8.9. It is not uncommon for scientists to estimate different magnitudes immediately after an earthquake.

The powerful earthquake also appears to have moved the main island of Japan by 8 feet (2.4 meters) and shifted the Earth on its axis.

A statement from Kenneth Hudnut, a geophysicist with the U.S. Geological Survey (USGS):

> At this point, we know that one GPS station moved (8 feet), and we have seen a map from GSI (Geospatial Information Authority) in Japan showing the pattern of shift over a large area is consistent with about that much shift of the land mass.

Reports from the National Institute of Geophysics and Volcanology in Italy estimated the 9.0-magnitude earthquake shifted the planet on its axis by nearly 4 inches (10 centimeters).

This seismic event also caused the largest earthquake shift ever observed directly. Japan's seabed shifted as much as 79 feet (24 meters) in an east-west direction.

David Applegate, a senior science adviser for earthquake and geologic hazards for the U.S. Geological Survey, said the quake ruptured a patch of the earth's crust 150 miles long and 50 miles across.

The powerful seismic activity was obviously already brewing prior to March 11th, but we shall entertain the idea that the physical plate tectonics could have been goaded on and augmented by both the M and X-class solar flares that occurred two and three days before the 9.0 earthquake.

The vast amount of speculation that surrounds any phenomenon can render it difficult to examine clearly. Some scientists have theorized that disturbances from the Sun that upset Earth's magnetic field may lead to instability in our planet's tectonic plates. If so, the mechanics of these proposed interactions are far from being well understood.

Seismologists are still stubbornly skeptical of these ideas because they remain unproven. Many of the Earth-Sun interactions are still beyond the current understandings of modern physics.

There are always times when the ideas of different scientific disciplines do not necessarily integrate well. How much do Seismologists understand about Astrophysics or Heliophysics? Electromagnetic currents flowing up from under the ground during a solar storm are measurable. How do those currents fit into theoretical models of contemporary Seismology?

Links between space weather and earthquakes have been down played by the United States Geological Service. Below is the official statement from the USGS:

> By Dr. Jeffrey Love
> Solar flares and magnetic storms belong to a set of phenomena known collectively as "space weather". Technological systems and the activities of modern civilization can be affected by changing space-weather conditions. However, it has never been demonstrated that there is a causal relationship between space weather and earthquakes. Indeed, over the course of the Sun's 11-year variable cycle, the occurrence of flares and magnetic storms waxes and wanes, but earthquakes occur without any such 11-year variability. Since earthquakes are driven by processes in the Earth's interior, they would occur even if solar flares and magnetic storms were to somehow cease occurring.

Is the USGS burying their head in the sand? The United States Geological Service is definitely brick & mortar, and Dr. Love *is* a doctor. The statement must be correct. No? Is the Earth flat? Let us examine this phenomenon closer.

The first axiom that we should apply to any given scientific "fact" is that they may not, *in fact,* be infallible. The history of scientific theory illustrates this very clearly. Psychologists could, and probably have had a field day endeavoring to explain the human tendency to calcify beliefs into the "known".

An excellent example of this quirk in human nature is the term *Newton's Laws.* These *Laws* are still taught as dogma in universities and high schools world-wide. In reality Isaac Newton's ideas may have been dynamic in his day, but they have always been *theories* that are now hundreds of years out-of-date.

Today we honor Newton for his theories which, at the time were quite revolutionary. However, we must not forget that in Newtons day, religions (plural) were very threatened by his ideas, while the common populace thought he was nuts--a total fruitcake.

Our history can be used as a mirror--allowing us to step back and get a better perspective towards our attitudes today. In the early chapters of the 21st century, society is still basing belief systems on Classical Physics, and are particularly wary of the ideas of the physicists of our own era.

Physicists do tend to be ahead of their time, and when they get together every several years for their international conferences they speak a unique language, or *Physicsese,* that most of us frankly do not understand.

An important topic that world-class physicists continually share ideas about is *gravitational theory.* **They do not totally understand what gravity is.** They have plenty of ideas, but none of the models are perfect. Newton's model explained the movements of large planetary bodies well, but breaks down when approaching the nuclear level.

In our present era physicists tend to get really jazzed up about what is known as *Quantum mechanics.*

The physicist and author Fred Allen Wolfe, in his book *Taking The Quantum Leap,* gives us a relatively easy-to-read linear history of physics from prior to Newtons day into the present era. Wolfe lays out the basics of theories from important physicists, how they built on each others ideas, and how those ideas have evolved.

Scientists are learning more every day about the relationship between heliophysics and geomagnetic energy. The bottom line is though; much of it remains not particularly well understood.

The American Geophysical Union has performed two studies concerning the relationship between solar activity and seismic events; one in 2007, and another in 2009.

We feel that it is only responsible to include the entire statement from the AGU Spring Meeting 2007, abstract #IN33A-03 by Jain, R:

> We present the study of 682 earthquakes of ¡Ý4.0 magnitude observed during January 1991 to January 2007 in the light of solar flares observed by GOES and SOXS missions in order to explore the possibility of any association between solar flares and earthquakes. Our investigation preliminarily shows that each earthquake under study was preceded by a solar flare of GOES importance B to X class by 10-100 hrs. However, each flare was not found followed by earthquake of magnitude ¡Ý4.0. We classified the earthquake events with respect to their magnitude and further attempted to look for their correlation with GOES importance class and delay time. We found that with the increasing importance of flares the delay in the onset of earthquake reduces. The critical X-ray intensity of the flare to be associated with earthquake is found to be ~10-6 Watts/m2. On the other hand no clear evidence could be established that higher importance flares precede high magnitude earthquakes. Our detailed study of 50 earthquakes associated with solar flares observed by SOXS mission and other wavebands revealed many interesting results such as the location of the flare on the Sun and the delay time in the earthquake and its magnitude. We propose a model explaining the charged particles accelerated during the solar flare and released in the space that undergone further acceleration by interplanetary shocks and produce the ring current in the earth's magnetosphere, which may enhance the process of tectonics plates motion abruptly at fault zones. It is further proposed that such sudden enhancement in the process of tectonic motion of plates in fault zones may increase abruptly the heat gradients on spatial (dT/dx) and temporal (dT/dt) scales responsible for earthquakes.

In common speak, what the AGU researchers are saying is that their data *suggests that* solar flares have an effect on seismic activity, but it is not totally conclusive.

The Russians are on the same page as the AGU concerning this issue. "Unfortunately, we're expecting more severe cataclysms which may lead to large-scale human losses and destruction," says Baku-based Professor Elchin Kakhalilov of the Global Network for the Forecasting of Earthquakes. "I'm talking about even a possible shift of the centers of our entire civilization."

According to NASA, researchers using NASA's fleet of five THEMIS spacecraft have discovered a form of space weather that packs the punch of an earthquake and plays a key role in sparking bright Aurora's. They call it a *spacequake.*

Spacequakes create a trembling in the Earth's magnetic field. It is felt most strongly in Earth orbit, but is not exclusive to space. The effects can reach all the way down to the surface of Earth itself.

"Magnetic reverberations have been detected at ground stations all around the globe, much like seismic detectors measure a large earthquake," says THEMIS principal investigator Vassilis Angelopoulos, a space physicist at the University of California, Los Angeles.

It's an apt analogy because "The total energy in a spacequake can rival that of a magnitude 5 or 6 earthquake," according to Evgeny Panov of the Space Research Institute in Austria. Panov is the main author of a paper reporting the results in the April 2010 issue of *Geophysical Research Letters* (GRL).

"In general, Earth's magnetic field lines can be thought of as rubber bands stretched taut by the solar wind, which is actually charged particles flowing in all directions from the sun", said study coauthor Vassilis Angelopoulos

Another entity that does elicit powerful geomagnetic influence on our planet is our moon.

Because of seismometers left on the moon by Apollo astronauts, we know that Earth's gravity will trigger moonquakes. Why is it such a stretch to consider that the moon can do the same to Earth?

The moon's effects on plate tectonics has been down-played by the USGS, whose official stance on the idea is negative. "There is still no known observation of an effect related to the moon and seismicity." (Bellini--USGS geophysycist)

The USGS claims that since the mass of the Earth is 82 times greater than the moon's mass, Terra's smaller neighbor enforces little influence on it. Good point.

Some historical data does however, support that solar and lunar influences can contribute to triggering catastrophic events including earthquakes and volcanoes.

There is a higher incidence of earthquake activity in the Northern Hemisphere when the moon is north of the Equator and an increase in earthquake activity in the Southern Hemisphere when the moon is south of the Equator.

In conclusion, the jury is still out concerning the potential that solar or lunar activity can play a part in the physical mechanisms that cause earthquakes. The majority of mainstream scientists give both possibilities a reverberating thumbs down.

Hard data tells us that earthquakes occur constantly in different areas on our planet without corresponding notable influences from the moon or the sun.

We *do* however urge readers to continue to observe events with an open mind concerning new scientific potentials. The correlation between extreme solar weather and the earthquakes off the coast of Japan in March of 2011 does leave a lingering, suspicious taste in our mouths that is difficult to so casually dismiss.

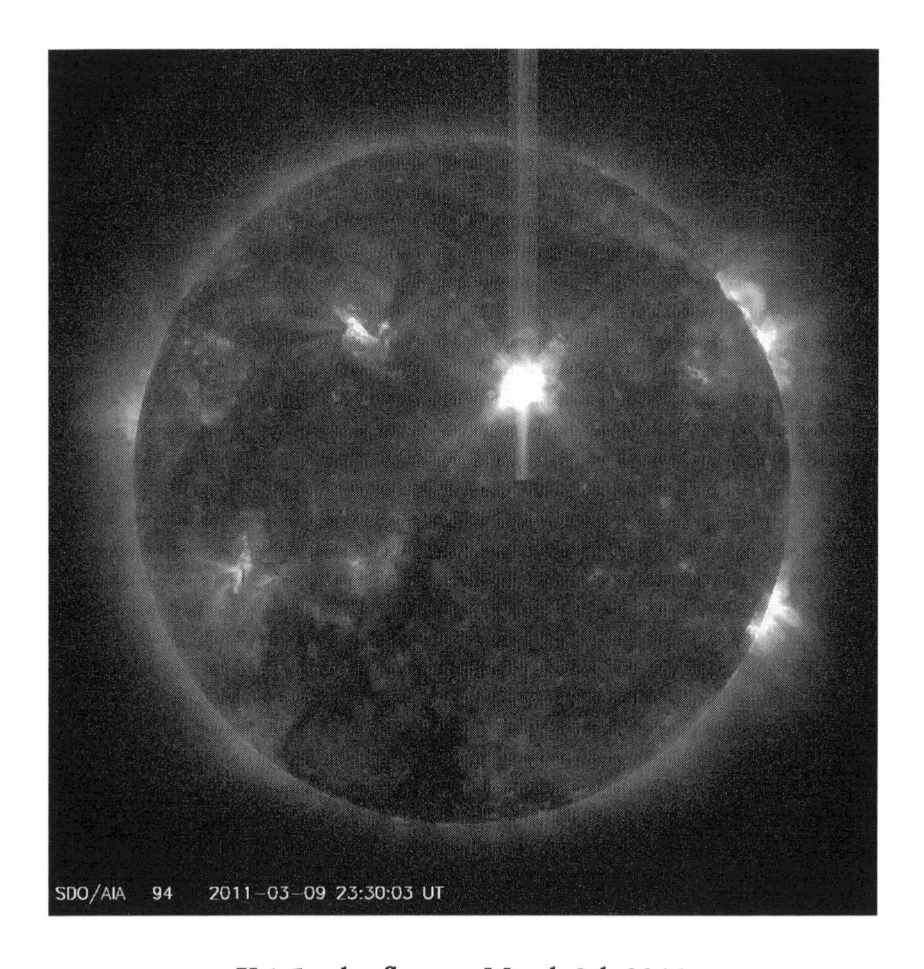

X 1.5 solar flare on March 9th 2011

This one, along with an M 3.7 the day earlier
preceded the Japanese earthquakes.

Courtesy of NASA

Recent Solar Activity & Seismic Activity Coincidence?

3/11/11 **9.0 earthquake off of Honshu, Japan**

March 2011	8th	9th	10th	11th
	M 3.7 Class Flare CME Launch >	X 1.5 Class Flare CME Launch > 7.2 Foreshock Quake Honshu Japan	5.8 quake Kills 25 In China	9.0 quake Honshu Japan

Over 700 average 5.5 magnitude earthquakes continued for one week off the coast of Japan beginning on March 9th

Recent Solar Activity & Seismic Activity Coincidence?

8/3/10 **6.4 magnitude earthquake in Papau New Guinea**

	C-3 Class Flare	CME Travels to Earth >	6.4 Quake New Guinea
August 2010	1st	2nd	3rd

The incident in August 2010 is interesting because even though the solar flare was only a C-3 class, the event was quite dramatic. The entire Earth-facing side of the Sun erupted in a series of visible solar phenomenon.
The timing for the following quake in New Guinea was perfect, considering that it took about two days for the Coronal Mass Ejection to reach the Earth's magnetosphere.

Recent Solar Activity & Seismic Activity Coincidence?

2/22/11 **6.3 earthquake rocks Christchurch New Zealand**

	M Class Flare CME Launch >	X-2.2 Class Flare CME Launch>	6.3 Quake New Zealand
February 2011	13th	14th	22nd

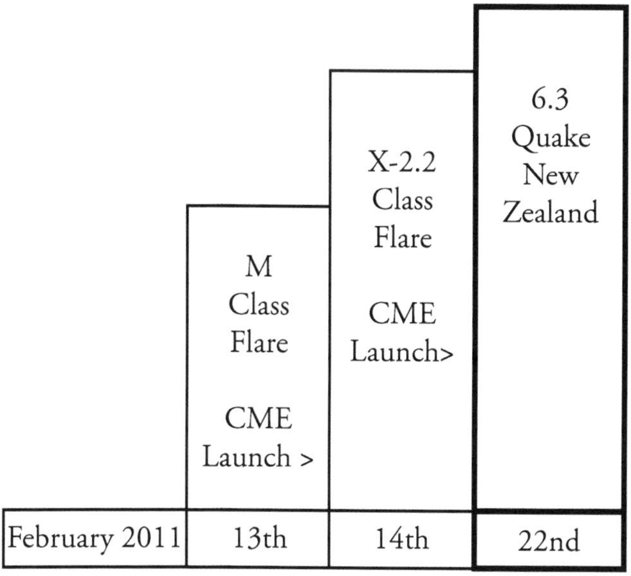

A full week elapsed between the X-2.2 class flare and the earthquake in Christchurch, New Zealand.
The CME struck Earth's magnetosphere on Feb. 18th.
It should be noted that this earthquake occurred during a particularly powerful lunar cycle.

EMP Devices

An electromagnetic pulse (EMP) device is a weapon designed to nullify electronic equipment.

Our attention in this treatise is focused on solar weather, but we cannot discuss Faraday shielding without at least outlining the basics of EMP devices.

One could only hope that mankind has evolved far enough socially to cease throwing stones at those in it's own tribe. However, recent sociopolitical developments are disturbing, and point to a continuance of primitive, violent, aggressive behavioral patterns.

If evolved ET's do exist, why should it not surprise us that we may have only limited interactions with them?

Large EMP devices require a detonation which is forced through a magnetic field to create a ramping current pulse to the tune of tens of hundreds of microseconds and peak currents of tens of millions of amps. The shock wave is created when a stream of highly energetic photons collides with atoms of low atomic numbers and cause them to eject a 'pulse' of electrons.

In military terminology, a nuclear bomb detonated hundreds of kilometers above the Earth's surface is called a *high-altitude electromagnetic pulse* (HEMP) device. A HEMP weapon is designed to create continent-wide mayhem.

The three components of a nuclear HEMP, as defined by the International Electrotechnical Commission (IEC), are called E1, E2 and E3. Each component of the weapon is designed to inflict damage in a unique fashion, like a one-two-three punch.

The E1 pulse is the fastest component of nuclear EMP; too fast in fact for ordinary lightning protectors to provide effective protection against it. In fact, the E1 pulse is designed to nullify electronic protection systems.

The E2 component is an "intermediate time" pulse that lasts from about 1 microsecond to 1 second after the beginning of the electromagnetic pulse. The E2 pulse is designed to destroy a wide range of electronic equipment.

The E3 component is a very slow pulse, lasting tens to hundreds of seconds. The E3 component has similarities to a geomagnetic storm caused by a powerful solar flare. Like a geomagnetic storm, E3 can produce geomagnetically induced currents in long electrical conductors, which can do further damage to components such as power line transformers and large equipment.

Militaries also have non-nuclear EMP devices, known as non-nuclear electromagnetic pulse (NNEMP) weapons, which are designed for small target strikes such as ships.

Unlike the Sun, which has no personal agenda to harm, EMP devices are more difficult to shield against because they have been designed specifically for a purpose. Radiation hardening built to withstand an attack by an EMP weapon requires considerably more robust shielding than that which is needed for protection from solar flares.

EMP's are designed to deliver a powerful punch (pulse) at very high frequencies, particularly concerning the E1 and E2 components. The brief spike in power drives the high frequency particles to such a degree that they can sneak through tiny holes in shielding that lower frequency particles from solar events cannot pass through.

What is particularly disturbing is the rumored presence of "Super EMP's". In some open-source writings, Russian and Chinese military scientists have described the basic principles of weapons designed to generate an enhanced EMP effect. The super EMP weapons are supposed to be able to destroy even the best protected U.S. military and civilian electronic systems.

Military secrets. Therefore, we do not know how much, if any amount of shielding can protect us from an EMP weapon.

Somehow that brings to mind the poignant scene from the popular movie *The Two Towers,* by J.R. Tolkien; when a massive, ugly army is breaking down the doors of the inner sanctum of Helm's Deep.

Preparing to fight to the death, the King of Rohan stops to ponder for a moment, and asks a profound question:

"What can men do against such reckless hate ?"

No one really knows if those evil little men will ever actually attack with their EMP weapons.

Our Sun however, will eventually throw a temper tantrum, causing havoc here on Earth. It is not a question of if--but when. It is simply a matter of time.

Michael Faraday

Born to humble parents in England in 1791 and lived until 1867. Although Faraday received limited formal education and knew little of higher mathematics, he was one of the most influential scientists in history. He has been acknowledged as one of the best experimentalists in the history of science.

Known most for his work with electromagnetism, he was also an accomplished chemist, having discovered benzene, and investigated the clathrate hydrate of chlorine. Faraday was also an expert in the formulation of high quality optical glass.

The Faraday Cage

In 1836, Michael Faraday observed that the charge on a charged conductor resided only on its exterior and had no influence on anything enclosed within it.

To demonstrate the idea, he built a metal shielded room and then proceeded to blast it with high-voltage discharges from an electrostatic generator. He used an electroscope (grandfather of a voltmeter) as a measuring device to demonstrate that there was no electric charge on the inside of the room's walls.

Although the cage effect has been attributed to Michael Faraday, it was actually Benjamin Franklin in 1755 who first observed the effect in a simple cork and can experiment. Later, Faraday's famous ice pail experiment was based on Franklin's observations.

The idea is to install electrical equipment inside a metallic enclosure to protect it against outside electrical influence, or the opposite; to contain certain frequencies emanated by electrical equipment installed inside an enclosure.

In many modern industrial applications, the protection device actually is a cage--hence it's name. Copper screened rooms surround sensitive test equipment--shielding the equipment from RF (radio frequencies) to ensure their accuracy. The holes in the screen only need to be small enough to block the frequencies with wavelengths larger than the openings. Since RF wave lengths are large, the holes may be a big as 20mm (3/4") or larger.

An example of a Faraday cage we are all familiar with is a common microwave oven, which is designed to trap microwave frequencies inside the device. The metal case provides the shield, as does the door, which has a continuous screen. We can see through the screened door because the 1.5mm holes in it are easily large enough for visible light waves to pass through. The microwave frequencies however, may not pass because they have bigger wavelengths of about 120mm.

Due to the wide range of wavelengths we are endeavoring to block we are favoring solid shielding, but we will still refer to the protective systems as "cages".

Shielding is taken very seriously by corporations, industry, and militaries. Holland Shielding Systems B.V. in the Netherlands produces electromagnetic shielding for scientific and medical instruments, military use, servers, and conference rooms (electronic eavesdropping).

U.S. government buildings have incorporated Faraday cage principals in their designs for decades. FEMA headquarters buildings incorporate dome-shaped earth-bermed structure designs, while under the buildings lays a copper mesh that extends out from the base and is secured by multiple grounding rods.

Many products are available on the wholesale and retail markets that are designed to protect personal equipment using various shielding techniques. Anti-static dissipative bags and foam products, mats and shielding tape are common products that are touted as providing some low-level protection from electromagnetic interference.

Most Faraday cage protective systems have not been tested in the 21st century against real-time electrojets or man-made EMP devices. The effects of EMP weapons on electronics and power systems was observed in the 1960's by both the U.S. military and the old Soviet Union, but the data is limited and electronic equipment was built more robust in that era.

Our Sun is a restless giant, and remains unpredictable even using our finest equipment and most talented heliophysicists. We must therefore acquiesce to recognizing our Sun as the proverbial wild card. We really do not know what kind of rabbit the Sun is going to pull out of its hat and when it will do so. That leads us to a key question: How much and what kind of shielding should we use?

One of the dangers we face is human nature itself. Some mature boy scouts and girl scouts will study and plan but not *do*, finding the task daunting. Others act without careful planning. We are urging everyone to do *something*, even if it is to only protect equipment deemed most useful or valuable.

A practical approach would be to make a list of electronic items that you would like to potentially protect, then rate them in order of importance. That old VCR that's rarely used anymore would be on the bottom of the list, while a ham radio might be considered top priority. Protect the most important gear by applying EMP-grade hardening techniques, then work your way down the list perhaps using more elementary shielding.

Basic Faraday cage design theory

A Faraday cage is a conductive enclosure used to block electrostatic and electromagnetic fields. The effectiveness of the device depends ultimately upon the material used, its thickness, the size of the shielded volume and the frequency of the fields of interest. Also important are the size, shape and orientation of apertures in a shield to an incident electromagnetic field, as well as the type of grounding used (if any).

A unique feature of a charged piece of metal is that, no matter what its shape is and if current is zero, then the electric field inside the piece of metal has to be zero. Free charges in the metal go to the surfaces of the metal and arrange themselves so that the electric field is zero everywhere inside the metal.

A Faraday cage operates on the same principle as a piece of metal. Both are conductors, but a Faraday cage could be described as a *hollow* conductor.

Electromagnetic radiation consists of coupled electric **and** magnetic fields. The **electric** field produces forces on the charge carriers (typically electrons) within the conductor. As soon as an electric field is applied to the surface of a Faraday cage, it induces a current that causes displacement of charge inside the conductor/cage that cancels the applied field inside, at which point the current stops.

Similarly, varying **magnetic** fields generate eddy currents that act to cancel the applied magnetic field. The result is that electromagnetic radiation is reflected from the surface of the conductor: internal fields stay inside, and external fields stay outside.

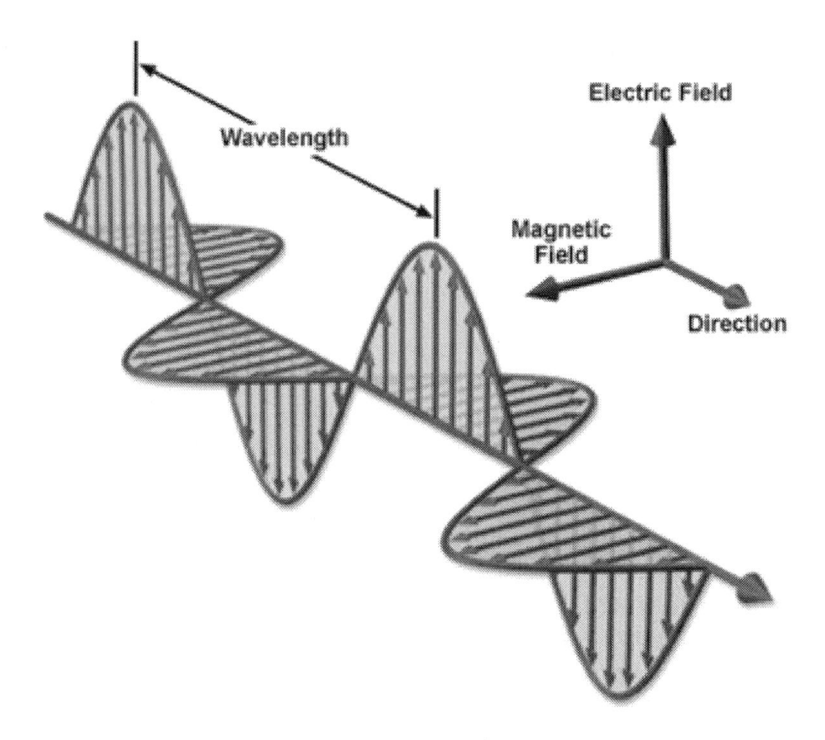

Electromagnetic waves are two waves operating in tandem. An **electrical** wave creating an electrical field moves along one plane, and a **magnetic** wave creating a magnetic field that moves perpendicular to the electrical field, as shown above. The two fields feed off each other, creating a self-propagating wave

Both of these fields are thwarted by the Faraday cage, but in different ways. The **electric** field is diverted by causing the electrons in the metal of the cage to rearrange, neutralizing any charge within the cage.

Faraday also illustrated this phenomenon with an ice pail. When **magnetic** waves come in contact with the Faraday cage, they create a current in the conductor known as an eddy current. These eddy currents, in turn, create magnetic fields that oppose the field of the oncoming waves. That effect causes those waves to be blocked from the interior of the cage.

The conductor (cage) does not respond to static magnetic fields unless the conductor is moving relative to the magnetic field. Faraday cages cannot block static and slowly varying magnetic fields, such as Earth's magnetic field. That is why a compass will still work inside of a Faraday cage.

An example of a conductor moving relative to a magnetic field would be a common electric generator. The *rotor* spins in the presence of the surrounding *stator*, one being magnetic, while the other is the conductor. The current generated by this arrangement varies depending on how fast the rotor is spinning.

Unlike a generator, a Faraday cage is a conductor, but it does not produce a current because the cage has no magnetic field moving near it. Externally applied electromagnetic fields (lets say from electrojets or EMP's] produce forces on the charge carriers (usually electrons) within the conductor, generating a current that rearranges the charges. Once the charges have rearranged so as to cancel the applied field inside, the current stops.

The primary principle behind the working of a conductor is the presence of free electrons and bound electrons.

It is the free electrons that determine the charge that is built up on the surface of a Faraday cage due to the presence of an external magnetic or electric field. Therefore, the charges re-accumulate in such a manner that they are all on the surface and so that the resultant field on the inside of the conductor is zero in magnitude.

Oftentimes the exterior of the Faraday cage is grounded so as to allow all current to pass through to the ground and retain the electrical neutrality of the conductor. This guarantee ensures the safe usage of the Faraday cage especially when it is used over spaces that contain electrical devices that would result in a space that is filled with electrostatic fields.

If the cage is grounded, the excess charges will go to the ground instead of the outer face, so the inner face and the inner charge will cancel each other out and the rest of the cage would remain neutral.

If the cage is not grounded it will function within limits dictated by its design and the materials used in it's construction.

We must therefore, ask another key question:

To ground or not to ground?

Concerning the grounding issues of Faraday cages, the world-wide web has become a virtual sparring ring for those who adhere to either the grounded or the ungrounded camp. After each round, the contenders brood in their respective corners, after verbally pommelling each other with a kind of ruthless abandon only observed in internet forums. All of the drama occurs devoid of considering other options. Somehow we are presented with only two choices: multiple choice "A" (grounded), and multiple choice "B" (ungrounded). We shall entertain the possibility of the existence of multiple choice "C".

Could both grounded and ungrounded systems function--each in their own ways? Perhaps we can create our shielding systems utilizing *both* ungrounded and grounded enclosures.

While the textbook answer is no, Faraday shields do not need grounding; in practise virtually all terrestrial-based formal Faraday cages used for industrial and medical purposes are well grounded. It seems the verbal contestants are: the academic theorists *versus* the practical industrialists. (A cartoonist could have fun with that one.)

Ungrounded static-dissipative bags and aluminum foil may work adequately to block low level radiation, but for high intensity protection an ungrounded cage needs to have sufficient wall thickness of a highly conductive material. The skin depth and the conductivity of the skin material is directly relative to the size of the enclosure. The larger the cage--the thicker the wall needs to be, and the more conductive the metal.

How well a Faraday cage functions depends on the frequency and level of intensity of radiation striking it. A grounded cage has the advantage of being able to sluff off excess energy into it's ground rod. An ungrounded cage will more easily loose efficiency as energy levels increase, eventually storing energy short-term as a capacitor, then re-radiating it as a conductor--both inside and outside of the enclosure.

The questions we have had concerning grounded systems have to do with the influence of GIC subterranean currents that can travel up into the cage via the ground rod. How much current would it take coming out of the ground to reduce the ground rod's effectiveness to zero? At what point do subterranean currents *induce* energy into a Faraday cage?

Due to the limits of real-life tests these questions remain unanswered. We are suggesting ideally putting ungrounded metal enclosures inside of a larger grounded cage. First we will study grounding options.

Grounding

The electrical ground in your home (via the ground lug of an electrical outlet) will not work well as a ground for a Faraday cage. It's connected to too large of a web of wire that can act as an antenna.

To illustrate this point; If an ungrounded input of an ungrounded oscilloscope is hooked up to the ground of an electrical outlet there will be so much "noise" present that it can actually charge a Faraday cage.

Part of the problem is that we're talking about a whole range of wavelengths here. A solar flare GIC will induce a (relatively) slow waveform rise, and so is more easily shielded against than a nuclear EMP device. Ham radio operators who work in the VHF or UHF ranges should be aware of this. A ground that is suitable at the 3.5Mhz frequency will be totally unacceptable at the 350Mhz frequency. The reason is because as the frequency of the signal goes up, the lengths of the ground paths become critical. At some point, they stop acting like grounds and start acting like antennas, and can re-radiate energy.

Ham radio antennas are generally too short to permit the build up of voltages large enough to do a lot of harm. Of course, if you have mile long rhombics or beverage antennas you may want to make it a habit to disconnect them. For those wires, a very good way to protect them is to install a grounded bracket, drill two large holes and screw in a couple of spark plugs. Adjust the spark plug gap so that there is just no arc at the highest power you transmit at. That will protect your radio in case you leave it connected. You can also install a choke coil to ground to bleed off any DC and static voltage buildup.

During a slow buildup of a voltage such as a solar storm or static electricity the chokes will essentially short any voltage to ground and therefore it will never get to a high enough level to fire the spark plugs. But long wire antennas pick up a voltage component during a thunderstorm even though the lightning could be miles away. Those wave forms are a lot faster than the wave form of a solar storm or static buildup. Lightning wave forms have a rise time fast enough to "ring" with the chokes and thus creating a high voltage spike. The spark plugs will fire and so effectively short this spike to ground via the arc. Once the energy is dissipated, the spark will extinguish and the antenna will see a high impedance again until the next lightning event.

As good as this system is for lightning , static build up, and solar events, it will not protect against an EMP where the E1 wave form is so fast the spark plug has no time to fire. In that situation all of the energy would be traveling to your radio. The EMP waveform is fast and powerful enough to ring any metallic structure--making them a transmitter on their own at the natural frequency of the structure. This has the effect that there is zero protection from it besides using fully shielded rooms or equipment with high voltage vacuum tube front ends.

The same paradigm applies to the length of the grounding cable that goes from the Faraday cage to the ground rod(s). Ideally the rod should be close to the enclosure, or it can begin to act as an antenna. On that same note, all equipment cables should be fully tucked into their respective cages because they could function as antennae as well.

Housing water pipes are electrically resonant therefore are not considered an adequate ground source.

We need a totally independent earth ground to function properly. The optimum would be a *dedicated* ground--one that is used solely for the purpose of the Faraday cage.

The military optimum for grounding is as follows: A heavy copper ground wire attached to a copper rod 9 feet long pounded into the ground and a 12 inch hole around the top of the ground rod filled with water. Salt is a great way to make water more conductive.

We may not all be able to live up to military standards. Driving a ground rod *9 feet* into the ground can be rather difficult or impossible in some areas with rocky soils. Some sources suggest bending the rod as an option, allowing it to be laid in a trench. Unfortunately, bending the rod reduces its conductivity, rendering it less effective. If you must bend--do it carefully using a large *slow* bend. Many newer state building codes in the U.S. require more than one ground rod for grounding electrical panels in homes. The rods are separated by several feet and connected to each other with heavy solid copper wire.

Ideally, use AWG #4 or thicker bare copper wire for large equipment. The shorter the ground wire the better--with no sharp bends. Sharp bends will cause resistance, lowering the effectiveness of the ground. The wire is attached to the rod with a special clamp that must be purchased separately. 5/8" inch diameter round rods are available anywhere home construction supplies are sold. Available choices are either copper plated or galvanized rods. The copper plated rods are recommended.

Professionals use roto-hammers with special tips for driving in ground rods. The rest of us must resort to manual driving methods, the typical tool of choice being a heavy two-handled metal fence post driver. All feed stores that stock fencing supplies sell metal fence post tools. Do not attempt to drive a ground rod in with a sledge hammer.

Some soil types, such as deep sandy dry soils provide poor grounding factors. In difficult situations, we can use what is termed *concrete encased ground electrodes.* Concrete poured around a ground rod will increase it's effectiveness. Cured concrete that is in contact with the earth is a good conductor, particularly if it is wet. The resistance factors of concrete are as follows:

Dry concrete above grade 1-5 M ohms
Dry concrete on grade 0.2-1 M ohms
Wet concrete on grade 1-5 k ohms

Also, the property of concrete being slightly alkaline causes steel to become passive with regard to corrosion when steel is embedded in concrete.

Research has been performed by companies endeavoring to formulate what is known as *conductive concrete.* Some (or all) of the aggregate (gravel) normally in concrete can be replaced by conductive materials, such as metal shavings or metal scraps. It has been suggested that conductive concrete could actually be used for Faraday cage housings. Steel fiber has been added to concrete to increase its strength, as well as carbon fiber for specialized applications. Like metal, carbon fibers possess an additional advantage of having a high electrical conductivity.

14 gauge copper wire bolted and soldered to a claw clamp

Even though solid ground connections are preferred, quick connectors can also be used. Better than standard alligator clips, 1 5/8" claw clamps are rated at 5 amps and are available at Radio Shack. These clamps allow for both mechanical and soldered connections.

The presence of carbon fibers or metal fibers will greatly increase the electrical conductivity of the cement.

Simple homemade conductive concrete formula:

1 shovelful of Portland cement
3 shovelfuls of washed sand
1-2 shovelfuls of metal scraps
Water

Make certain that you mix your dry ingredients well first before adding any water.

It would be ridiculous to use 1/4" thick #4 wire to ground an anti-static bag. For lightweight cages, use 12 or 14 gauge stranded or solid copper wire. Solid copper wire can be purchased by the foot at big box stores, and stranded wire is available at automotive parts outlets. Feel free to use more than one ground wire.

Heavy duty copper grounding wire sets with molded, crimped lugs can also be purchased at automotive supply outlets. These are designed to ground engines to frames, but will work well for our purposes. The lugs will accept 5/16 bolts, and the wire is the flexible stranded type.

Electrical connections 101

In our modern era there are many who have never soldered, so please forgive our redundancy in stressing these points. Your electrical connections could very well be the weak link in your systems. It is a good idea to over-perform when fabricating your grounding connections.

Whether you are using mechanical and/or soldered connections make sure that the mating surfaces are freshly ground, shiny metal. 60, 80, or 100 grit sandpaper works well for this purpose.

Soldered connections can be done with a soldering iron or a propane torch. Use electrical (conductive) solder--**not** plumbing solder. The key is to get the **parts** hot enough to melt the solder to flow into the joint. If using a propane torch be careful not to over-heat the parts. Solder is designed to work well within a temperature range.

You will struggle less when soldering small parts if they are mechanically connected first. Bolt or screw parts together and *then* solder them.

Faraday shielding wall thickness

The rule of thumb is--the thicker the better, especially with less conductive materials. When high frequency radiation strikes a shield, some of it is absorbed by the skin, with the resulting electromagnetic cancellation called the *skin effect.* A measure for the depth to which radiation can penetrate the shield is termed the *skin depth.*

Skin depth requirements have been well researched by spacecraft designers, and are an important factor in their shielding. Like automobiles and aircraft, spacecraft are obviously ungrounded systems, and are often referred to as *internally grounded.* Thicker hulls provide more shielding effect in ungrounded craft.

Very sophisticated laminated shielding is used in satellite-based particle detectors called *graded-Z* shielding. Graded-Z shielding is a laminate of several materials with different Z values (atomic numbers) designed to protect against ionizing radiation. Compared to single-material shielding, the same mass of graded-Z shielding has been shown to reduce electron penetration over 60%.

Designs vary, but typically involve a gradient from high-Z (usually tantalum) through successively lower-Z elements such as tin, steel, and copper, usually ending with aluminium. Sometimes even lighter materials such as polypropylene or boron carbide are used. Some designs also include an outer layer of aluminium, which may simply be the skin of the satellite.

As homeowners, we do not generally have exotic high-Z elements available to us, but it is interesting to note the extent at which satellite builders are willing to go to protect their equipment.

A single layer of heavy aluminum foil will not be thick enough to protect equipment from an EMP attack, but could work sufficiently against most geomagnetic activity. It may be wise, if using aluminum foil, to design cages with multiple layers of it.

Shielding materials

Metals with high electrical conductivity make the best Faraday shields. In order of conductivity, common metals include gold, silver, copper, aluminum, and steel.

Using gold or silver for small shields may not be as outrageous as you think. Silver or gold plated copper or brass serving bowls and plates are highly conductive. A cell phone wrapped in anti-static foam and placed on a silver plate with a silver bowl turned upside down on it will probably protect the telephone from a solar flare.

Using that same theme, we can be very creative in our selection of objects to be used for, or retrofitted to become Faraday cages. We have included a short list of common objects that could be easily used in Faraday cage systems:

Steel shipping containers
Aluminum semi-truck containers
Metal-clad buildings
Steel file cabinets
Wood burning stoves
Steel office desks
Galvanized metal trash cans
Major appliances
Safes
Microwave ovens
Aluminum pressure cookers
Cooking stock pots with lids
Aluminum gun cases
Steel ammo cans

My favorite objects from the list above are the safe and the pressure cooker. The safe is a no-brainer, while the pressure cooker, being cast out of heavy aluminum, could easily be made into a small EMP-grade Faraday cage.

I find the notion of *building* a box out of wood and then proceeding to cover it with aluminum foil absurd. This has been suggested on the internet. Boxes and objects are commonly available.

If, however, a formal shielded room is needed, it can certainly be constructed with a wooden frame and clad with copper or aluminum sheeting. In medical facilities, lead sheeting or special lead drywall is used in walls surrounding X-ray equipment,

The key is to use a metal enclosure and eliminate or minimize any openings. The closer the surrounding container comes to a continuous metal skin the more protection will be provided.

Any holes in the shield or mesh must be significantly smaller than the wavelength of the radiation that is being kept out, or the enclosure will not effectively approximate an unbroken conducting surface.

Any breaks in the cage will cause gaps that allow for penetration by outside electromagnetic (EM) fields. For a mesh, or a hole drilled into a solid box, the penetration of EM radiation is limited to oscillations that have wavelengths shorter than 2 times the diameter of the opening. Therefore, a 1 inch opening would allow 2 inch and shorter wavelengths to pass through it.

Available Faraday Shielding Products

A list of companies that specialize in shielding products is provided in the Resources section of this book. Complete Faraday cages designed for industrial use can be purchased, as well as Faraday tents, boxes, racks, foils, mesh, cloth, tapes, gaskets, conductive adhesives, and conductive paints.

Static dissipative shielding bags are very common and available on Amazon. They will not provide enough shielding for our purposes by themselves, but are certainly usable in nested Faraday systems. They feature insulative layers on both the inside and outside of the bags, but we feel more comfortable using a bit of additional insulation around the electronic device requiring protection before putting it in the bag.

Typical construction for static shielding bags is either 3 or 4 layers of thin materials using what they call *buried metal construction*; a conductive aluminum shield layer is buried between static dissipative polymeric plastic layers. Typical specs: Resistance: 1 x 105 to 1 x 1012 ohms/sq., Static shielding: <30V ; Static decay: <0.05 sec; Tensile strength: >25 lb; Light transmission: 40%; Thickness: 3.1 MILS.

Metallurgic Bonding

For nuclear EMP-grade radiation hardening, adequate metallurgic bonds need to be created between all connecting surfaces. For protection against solar-induced GIC's however, that level of hardening may not be necessary.

We will use an ammo box/Faraday cage as an example to illustrate some of the properties of metallurgic bonding.

Suppose you've meticulously cleaned off all of the paint and dirt around where the lid seals onto the ammo box, and you've used a wire brush or sandpaper to make it shiny. When you seal up the box, your ohmmeter will show a good connection using direct current. If you measure the connection with ordinary AC household current (between 50 and 60 hertz), you'll see it's also a very low inductance connection, with practically no energy lost between the two sides of the connection. However, as you increase the frequency of the current, the inductance will change. Remember, there's no real metallurgic bond between the two sides of the lid, only a physical connection. This means there's going to be both capacitive and inductive effects between the two sides of the connection. These effects will increase as the frequency increases, and at some point, a significant amount of energy will no longer be transferred through the connection, but will be reflected back to the source. At this point, you have a tuned circuit radiating energy. This energy can be re-radiated inside the shield, and could be powerful enough to destroy the electronics stored within.

If you have a metallurgic bond (solder or weld) between the two sides, then that minimizes the inductive and capacitive effects but will not eliminate them. Even though it's small, as the frequency goes up the electrical break formed by the solder or weld will again become significant and high frequency energy will be re-radiated inside the shield.

It should be noted that both the capacitive and the inductive potentials of Faraday systems are much more problematic with ungrounded cages.

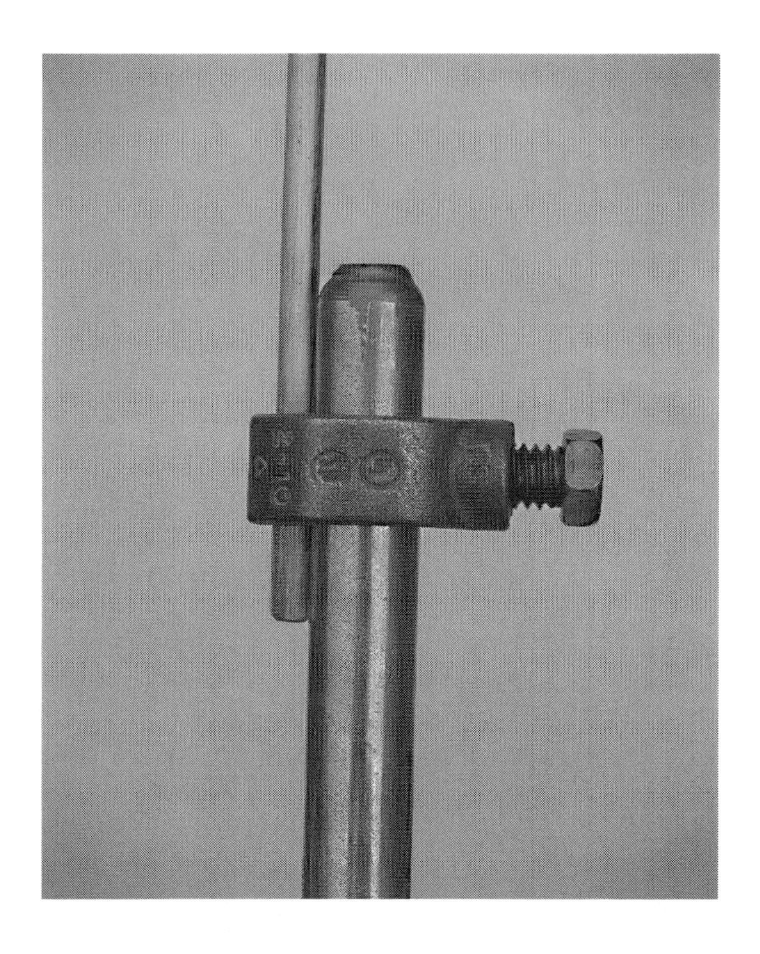

The end of an 8 foot, 5/8" diameter copper-plated ground rod attached to a 4 gauge solid copper wire with a brass clamp. These are commonly available parts in construction outlets in the USA.

Photo is actual size

Layering & Nesting

Using different layers of protection to create Faraday cage *systems* is highly recommended, particularly if one has less than optimum grounding or external shield materials.

An example of this technique would be to put each small individual electronic device in anti-static foam sheeting--then in a static-dissipative bag or foil wrap--followed with another layer of insulative sheeting. The triple wrapped devices would then be placed inside of an ammo can, which itself is placed on a non-conductive surface (like wood)--all inside the larger grounded Faraday enclosure. The static-dissipative bags and the ammo can are not grounded, while the mother cage *is* grounded.

This is where Faraday cage builders can be very ingenious in their selection of practical, available objects for use inside the larger cage. Copper foil and copper cooking pots (with lids) become practical, while aluminum, stainless steel and cast iron are commonplace.

Thrift stores like Goodwill and Salvation Army thus become the official shopping market for cheap Faraday cage materials. Before you go shopping however, take a few minutes to peruse through your own kitchen cabinets and explore further into your garage. You may have to reassure your wife that you will not drill holes to install grounding lugs in her silver Revere ware. The objects inside the mother cage need not be harmed.

It is very important when creating nested systems that each layer is insulated from other layers, and each electronic device is insulated from all layers. Dry wood, plastic, cardboard, glass, styrofoam, and foam sheeting are all non-conductive materials that can be used as insulation between components and layers.

Automobile protection

Our cars and trucks are very important, expensive, and necessary pieces of hardware that definitely warrant special attention.

I has been suggested that if you own a car built between 1974 and 1992 you can buy an extra coil, computer module, and (for EMP) an spare alternator. Those parts could be stored in a small Faraday cage (like a pressure cooker) in the car itself, along with the tools necessary to install them.

Metal bodied cars are already incomplete Faraday cages. The shielding can be improved by laying a shielding mat with a ground wire on the dashboard and clipping the wire to the inside of the metal dashboard where a fresh metal surface has been exposed. The ashtray is an option. Install a ground rod where you park the car and connect it to the car frame or metal bumper with both leads of a jumper-cable set.

Faraday tents sized specifically for autos can be purchased and used for parked cars.

Typical pole barn construction style for barns, shops, and garages use steel roofing, siding, and doors. These buildings should provide moderate protection from most space weather if they are sufficiently grounded. Install ground wires to connect metal roofs to metal siding for each vertical wall, and proper buried ground rods.

Metal roll-up garage-type doors and swing entry doors are usually mounted to wooden (2x6) door bucks, so will require grounding. Ground the track of a roll-up door. Hinged metal doors may need a flexible, stranded copper ground wire attached to the face of the door itself, on the hinge side of the door. Grounding the hinges will not necessarily ground the door because the hinges are often mounted with wood screws into a wooden support in the door itself. Some all-steel doors, including commercial-types use machine screws to attach the hinges to the door. On those types of doors grounding the hinges should work. Please remember that windows provide zero protecting from radiation.

Metal buildings would probably not provide adequate protection for electronics from an EMP event.

Automobiles built before about 1974 that do not feature computer modules are less susceptible to GIC and EMP events.

The idea that American cars and trucks built before 1974 would still function after and EMP attack may not be accurate. Manufacturers switched from generators to alternators in the early 1960's. Any car with an alternator contains SS diodes within that alternator. However, those diodes generally have very high current ratings, on the order of several hundred peak amps or more, therefore may survive a GIC, but possibly not an EMP weapon.

Cars and trucks with ignition systems typical of the 1950's era would have a better chance of running after an EMP attack. That is, cars and trucks with generators instead of alternators. American pickup trucks were the last to switch over to alternators.

Concrete & Stone

Stone and concrete provide radiation shielding naturally due to the density of the materials, with stone weighing in heavier than concrete.

Underground shelters are becoming more common with the increasing awareness of the issues outlined here. 24 inches of material should be enough to protect yourself from extreme space weather. Once again, we still do not know how far under ground one would need to be to fully escape from an EMP attack, considering the ultra-secret nature of those devices.

Subterranean dwellers and homeowners with basements can use these factors to their advantage. A corner in your basement would be a good location for the placement of a Faraday cage.

Even though hollow concrete blocks are not made of particularly dense material, they are available and easily stacked. They may be filled with concrete. If you do choose to pour concrete into the blocks, mixing it slightly loose (with more water), will encourage it to more easily flow down into the matrix.

Please keep in mind that concrete, stone, and brick are conductive materials. Anything, or anyone you want to protect from the effects of radiation should be placed on an insulating layer--like wood.

Brick houses and buildings absorb a certain amount of radiation. Bricks are not very dense, but can offer a good first line of defense.

If you enjoy laying stone with mortar (like I do), stone walls can be made outside of a Faraday encloser.

Radiation deflecting roof designs are more difficult. If you can afford it, a copper roof would be ideal. Aluminum or steel roofing panels are an obvious choice. Genuine roofing tile (not fiberglass) and old slate roofs will provide modest levels of protection.

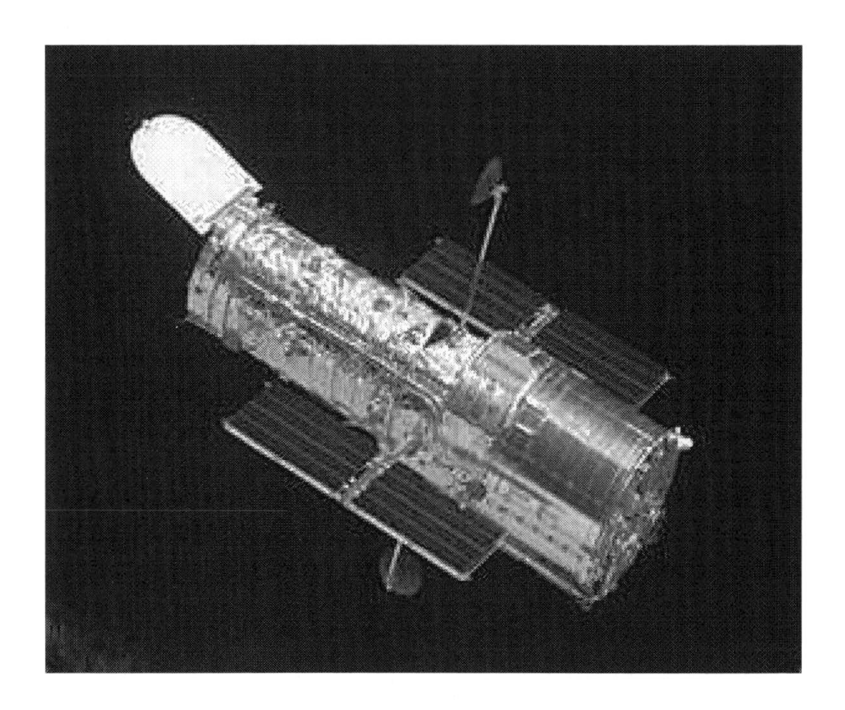

The Hubble Telescope
Courtesy of NASA

Watching The Sun

Since the 183 pound, beach-ball sized Sputnik 1 was launched in 1957, space agencies have put devices in orbit specifically to study the Sun. Dozens of spacecraft dedicated to heliophysics have lived out their lives unheralded by most of us, except as points of light quietly moving through the clear night sky.

Along with various ground based systems, today there are over sixteen satellites in orbit--each with their own special mission--observing our Sun, it's effects on the solar system, and inevitably--our planet. NASA calls this fleet of spacecraft the *Heliophysics System Observatory* (HSO). Each satellite focuses on a different aspect of heliophysics, has its own team of scientists and its own website. Some of the older spacecraft may, at any time cease to function, but below is a list of the currently operating satellites in the HSO fleet.

From NASA:

ACE / 8/27/97 Advanced Composition Explorer

ACE's mission observes particles of solar, interplanetary, interstellar, and galactic origins, spanning the energy range from solar wind ions to galactic cosmic ray nuclei.

AIM / 4/25/07 Aeronomy of Ice in the Mesosphere

Fifty miles above the ground, Earth's highest clouds form an icy membrane at the edge of the atmosphere. AIM's mission is to study these mysterious Polar Mesospheric Clouds.

CINDI/ 4/16/08 Coupled Ion-Neutral Dynamics Investigations

CINDI's mission is to understand the dynamics of the Earth's ionosphere.

Cluster / 7/16/00

The Cluster mission is an in-situ investigation of the Earth's magnetosphere using four identical spacecraft simultaneously.

Geotail / 7/24/92

The Geotail mission's primary objective is to study the dynamics of the Earth's magnetotail over a wide range of distance, extending from the near-Earth region to the distant tail.

Hinode (Solar-B) / 9/23/06

Hinode is exploring the magnetic fields of the Sun and is improving our understanding of the mechanisms that power the solar atmosphere and drive solar eruptions

IBEX / 10/19/08 Interstellar Boundary Explorer

IBEX is the first mission designed to detect the edge of the Solar System.

RHESSI / 2/5/02 Reuven Ramaty High Energy Spectroscope Imager

RHESSI's mission is to explore the basic physics of particle acceleration and energy release in solar flares

SDO / 2/11/10 Solar Dynamics Observatory

Solar Dynamics Observatory's (SDO) mission is to understand the Sun's influence on Earth and Near-Earth space by observing the solar atmosphere on small scales of space and time and in many wavelengths

SoHO / 12/2/95 Solar and Heliospheric Observatory

SoHO's mission is to study the Sun, from its deep core to the outer corona, and the solar wind.

STEREO / 10/25/06 Solar Terrestrial Relations Observatory
The STEREO mission employs two nearly identical space-based observatories - one ahead of Earth in its orbit around the Sun, the other trailing behind - to provide the first-ever stereoscopic measurements to study the Sun and the nature of its coronal mass ejections (CMEs).

THEMIS / 2/17/07 Time History of Events and Macroscale Interactions during Substorms
THEMIS answers longstanding fundamental questions concerning the nature of the substorm instabilities that abruptly and explosively release solar wind energy stored within the Earth's magnetotail.

TIMED / 12/7/01 Thermosphere, Ionosphere, Mesosphere Energetics and Dynamics
TIMED explores the Earth's Mesosphere and Lower Thermosphere (60-180 kilometers up), the least explored and understood region of our atmosphere.

TWINS / 3/13/08 Two Wide-angle Imaging Neutral-atom Spectrometers
The identical TWINS-A and TWINS-B observatories provide a new capability for stereoscopically imaging the magnetosphere.

Voyager / 9/5/77
The twin spacecraft Voyager 1 and Voyager 2 were re-designated in 1990 as the Voyager Interstellar Mission (VIM), to observe the edge of the solar system and interstellar space.

Wind / 11/1/94
Wind's mission is to measure crucial properties of the solar wind before it impacts the Earth's magnetic field and alters the Earth's space environment and upper atmosphere in a direct manner.

Image Courtesy of NASA

Recently, in early 2011 the two STEREO spacecraft finally were positioned on opposite sides of the Sun, giving us a 360-degree panoramic view of solar activity for the first time in history. Scientists have been able to observe all parts of the Sun, but not at the same time. This omnipotent view will help give us for a more complete understanding of heliophysics and allow for better space weather predictions.

NASA's Solar Dynamics Observatory (SDO) is the new crown jewel in their fleet of spacecraft observing the Sun. SDO was launched aboard an Atlas V rocket on Feb. 11, 2010, and is the most advanced solar observatory in the sky. The mission is the cornerstone of a NASA science program called Living With a Star (LWS). The goal of the LWS Program is to develop the scientific understanding necessary to address those aspects of the Sun and solar system that directly affect life and society.

We can benefit from the work being done by NASA and its teams of scientists in the HSO fleet.

Our first observational line of defense is ground based and space born telescopes, which monitor visual solar activity in various different light spectrums. A potentially powerful solar event is first observed and recorded.

Almost eight minutes after an event occurs on the Sun, x-rays and fast moving particles reach Earth, and are detected by satellites that are specifically designed to monitor them.

The influx of x-rays into Earths system that are powerful enough to be deemed an SID (Sudden Ionospheric Disturbance) will generate an alert that is important for us to be aware of. An SID may be the precursor of a much more powerful and damaging Coronal Mass Ejection. As we now know, the ensuing CME may take from 18 hours to 4 days to reach Earth. That gives us some time to prepare if the forecasters are predicting a potentially damaging CME.

Some good news is that a team at Stanford University in California, say they have a developed a technique that could give advance warning of the formation of sunspots before they become visible on the Sun's surface. The Stanford team used a novel technique called helioseismology, which is based on analysis of vibrations on the solar surface. The team discovered that these acoustic signals causing the vibrations moved faster in regions where sunspots were forming up to 65,000 kilometers deep. The resulting sunspots appeared on the surface between one and two days after the differences in vibrations were detected. The technique can also help predict the size and strength of the sunspot.

The Russian Academy of Sciences have a new project called *Interheliozond* that has been accepted into Russia's Federal Space Program. They plan on sending a probe very close to the Sun.
"This is going to be like a thermometer which would fly close to the Sun, measure its temperature, density and magnetic fields," Dr. Sergey Bogachaev, an Interheliozond engineer from Moscow. "We're making it from scratch: no one has ever done this before."

Receive Alerts!

At spaceweather.com one can sign up to receive free email alerts concerning various kinds of space weather.
From spaceweatherphone.com you can get alerts on your cellphone and email for $4.95 per month. You can choose what kind of alerts you want to favor, including: Geomagnetic storms, X-class solar flares, CMEs and solar wind gusts, solar radiation storms, and the interplanetary magnetic field.
When they get the data from the plethora of satellites and ground equipment dedicated towards the Suns activities, information is forward to subscribers.
Also, at the spaceweather.com website, the current information concerning the Sun's activities data is updated constantly. If this is not enough frighten or agitate, they also have a handy chart of "recent & upcoming Earth-asteroid encounters".
Or you can choose 3DSun, developed in collaboration with NASA scientists. When an alert comes into your cell phone it appears as a current image of the celestial body in question, originating from NASA's two STEREO satellites. Subscribers can view the Sun from different angles by rotating it on their screen, or zoom in to see it closer.

More websites from which you can receive data

NOAA/NWS hosts spaceweather.org, where you can quickly get data about current conditions from many locations on the planet. The data in the US originates from Boulder, Colorado.
NOAA official site www.noaawatch.gov/themes/space.php
www.solarcycle24.com
Utah State University hosts spaceweather.usu.edu
From the Canadian Weather Service www.spaceweather.gc.ca
South African data www.spaceweather.co.za
Ireland www.SolarMonitor.org
The European Space Weather Portal www.spaceweather.eu
Australia www.ips.gov.au/Space_Weather
Solar Terrestrial Dispatch www.spacew.com
ESA,s excellent site esa-spaceweather.ne
Follow space weather on Twitter www.spaceref.com/spaceweather
Track effects on GPS systems www.spaceweather.gc.ca/se-gps-eng.php
From Washington D.C. www.windows2universe.org/spaceweather
Spain Magnetosphere Universidad de Alcalá www.spaceweather.es
Science from US dept. of Commerce www.spaceweather.sflorg.com
University of Helwan/Egypt www.spaceweather-eg.org/sws
Universidad de Malaga spaceweather.uma.es
Space weather on Twitter twitter.com/spaceweather
Harvard hea-www.harvard.edu/.../SpaceWeather

Understanding Alerts

Solar flares are classified as A, B, C, M, X and X+ according to their intensity. "A" class flares are the weakest, while "X+" class are the most powerful. Each category has nine subdivisions ranging from, e.g., C1 to C9, M1 to M9, and X1 to X9, while X+ class are X10 and above.

NOAA has simplified the X-class for their purposes by referring to X class flares from 1 to 20 (and above). X1 through X20. This will most likely become the standard.

M class flares can result in temporary radio blackouts on Earth, but it is the X-class events that we are primarily concerned with.

The most popular index scientists use for giving a quick reading as to the level of geomagnetic intensity reaching the Earth is the *Kp* index devised by Julius Bartels in 1932. The Kp index is semi-logarithmic, and is from 1 to 10 . A Kp = 9 storm is roughly five times stronger than a Kp = 6 storm. On an ordinary day Earth's magnetic field weighs in from about 1.0 to 3.0 Kp. Kp values between 4.5 and 5.5 are classified as small storms, while storms between 5.5 and 7.5 Kp are considered large. Events eliciting values greater than 7.5 can result in obvious damage, like the 9.3 Kp event that induced the Quebec blackout in 1989.

Time is expressed in space weather reports as "UT", or "UTC", both acronyms for *Universal Time.* In 1928 the term Universal Time was adopted internationally as a more precise term than Greenwich Mean Time. Universal Time (UT) is a timescale based on the rotation of the Earth. The expression "Universal Time" is somewhat ambiguous, as there are several versions of it, the most commonly used being UTC, UT1, and UTO. All of these versions of UT are based on the rotation of the earth compared to distant celestial objects (stars and quasars), but with a scaling factor and other adjustments to make them closer to solar time.

NOAA has introduced the **Space Weather Scales** as a way to communicate to the general public the current and future space weather conditions and their possible effects on people and systems.

Geomagnetic Storms: disturbances in the geomagnetic field caused by gusts in the solar wind that blows by Earth

Scale	Level	Impact	Kp	Frequency
G1	Minor	Radio interference	5	1700 per cycle
G2	Moderate	Power grid disturbances	6	600 per cycle
G3	Strong	Some grid problems	7	200 per cycle
G4	Severe	Transformer trips	8	100 per cycle
G5	Extreme	Blackouts	9	4 per cycle

Solar Radiation Storms: elevated levels of radiation that occur when the numbers of energetic particles increase. Measured in Flux level of >= 10 MeV particles (ions)*

Scale	Level	Impact on Bodies	Flux	Frequency
S1	Minor	None	10	50 per cycle
S2	Moderate	Some	10^2	25 per cycle
S3	Strong	Elevated Risk	10^3	10 per cycle
S4	Severe	High Risk	10^4	3 per cycle
S5	Extreme	Very High Risk	10^5	1 per cycle

Radio Blackouts: disturbances of the ionosphere caused by X-ray emissions from the Sun. Measured using GOES X-ray peak brightness by class

Scale	Level	Blackout Effects	Class	Frequency
R1	Minor	HF Disturbance	M1	1 per cycle
R2	Moderate	HF Interuptions	M5	8 per cycle
R3	Strong	1 hour Blackouts	X1	175 per cycle
R4	Severe	2 hr. + Navigation Issues	X10	350 per cycle
R5	Extreme	Major HF & LF Issues	X20	2000 per cycle

Bibliography

Aguirre, E. L. (2005 May). Amateurs detect effects on earth. Sky and Telescope, 109. Available through ProQuest

Contreira, D., Rodrigues F. S., Makita K., and Brum, C. G. (2004). An experiment to study solar flare effects on radio-communication signals. Advances in Space Research, 36(12), 2455-2459.

Davies, Kenneth (1990). Ionospheric Radio. London: Peter Peregrinus Ltd.

The Effect of Solar Flares on the VLF Radio Waves Transmitted in the Ionosphere, a PowerPoint presentation by Sharad Khanal, 2004.

Leary, W. E. (2006, March 6). Scientists say next solar cycle will be stronger but delayed. New York Times.

Peter N., (2003, March 3). In Iraq, solar storms play havoc with communication. Christian Science Monitor, p. 15.

Radio Frequency Wikipedia Foundation, Inc.

The Science Behind the Monitors (Solar Center SID Site)

D. Scherrer, M. Cohen, To. Hoeksema, U. Inan, R. Mitchell, P. Scherrer "Distributing space weather monitoring instruments and educational materials worldwide for IHY 2007: The AWESOME and SID project" Advances in Space Research (a COSPAR publication) 42 (2008), 1777-1785.

VLF Remote Sensing of the Lower Ionosphere: Solar Flares, Electron Precipitation, Sprites, and Giant gamma-ray Flares (pdf)
(an IHY presentation given at the Fall 2005 AGU)

Stanford VLF Remote Sensing (pdf)
(a presentation given at Goddard Space Flight Center)

AWESOME Science Introduction (pdf)

Cohen, M. B., U. S. Inan, E. W. Paschal (2010),
Sensitive broadband ELF/VLF radio reception with the AWESOME instrument, IEEE Trans. Geosc. Remote Sensing, Vol 48, Issue 1, Pages 3-17, doi:10.1109/TGRS.2009.2028334.

Harriman, S. K., E. W. Paschal, U. S. Inan (2010),
Magnetic Sensor Design for Femtotesla Low-Frequency Signals, IEEE Trans. Geosc. Remote Sensing, Vol 48, Issue 1, Pages 396-402, doi:10.1109/TGRS.2009.2027694

Davies, K., "Ionospheric Radio," IEE
Electromagnetic Waves series 31, Peter
Peregrinus Ltd., London, UK, 1989.

Hargreaves, J.K., "The solar-terrestrial
environment," Cambridge Atmospheric
And Space Sciences series, Cambridge
Univ. Press, Cambridge, UK, 1992.

Jursa, Adolph S., ed., Handbook of Geophysics
and the Space Environment, Air
Force Geophysics Lab, Hanscom AFB,

MA, 1985.
Kelley, Michael C., "The Earth's Ionosphere:
Plasma Physics and Electrodynamics,"
Academic Press, 1989.
Sten Odenwald, The 23rd Cycle, 2003 Columbia University Press
Geophysical Research Letters Vol. 35 L20109, doj:10.1029/2008GLO35542
Discovery News August 25th 2010 "Is the Sun Emitting a Mystery Particle?", by Ian O'Neil
Chung Yu-Liu ,IEEE AESS Systems magazine, September 22, 2002 , A Study of Flight-Critical Computer System Recovery from Space Radiation-Induced Error

Kappenman, Hearing, Subcommittee on Environment, Technology, and Standards. Committee on Science, House of Representatives, 108th Congress, 10/30/03 Serial No. 108-31

Lawrence E Joseph, Apocalypse 2012, Morgan Road Books 2007

Resources

www.DirectMetals.com
www.westernrubber.com
www.LeaderTechInc.com
www.radiationshieldingsolutions.com
www.thermospray.com
www.omegashielding.com
www.glasscityplastics.com
www.captorcorp.com
www.vitatech.net
www.omegashielding.com
www.eride-electronics.com
www.BasicCopper.com
www.EMFBlues.com
www.visionteksystems.co.uk/emiclareglass.htm

Solar Flare Survival was formatted by Mark Stanley using InDesign CS5, and typeset in 12 point Adobe Garamond Pro. The display face is also Adobe Garamond Pro.
Printed in the United States of America.

Made in the USA
Lexington, KY
14 December 2011